*Notified in A.C.Is. 20th May, 1944.*

226
Publications
52

**RESTRICTED**

The information given in this document is not to be communicated, either directly or indirectly, to the Press or to any person not authorized to receive it.

# HANDBOOK OF ENEMY AMMUNITION

## PAMPHLET No. 10

**GERMAN MINES AND AMMUNITION FOR GUNS AND HOWITZERS.**

**ITALIAN GUN AND HOWITZER AMMUNITION.**

**JAPANESE 25 M.M. H.E. CARTRIDGE.**

*By Command of the Army Council,*

THE WAR OFFICE,
    20th May, 1944.

# HANDBOOK
## OF
# ENEMY AMMUNITION

## CONTENTS TABLE

### German Ammunition

|  | Page |
|---|---|
| Ammunition markings and nomenclature ... ... ... ... | 1 |
| S-mine (S.Mi.35) ... ... ... ... ... ... | 4 |
| Igniter S.Mi.Z.35 for S-mine ... ... ... ... | 7 |
| Wooden box mine 42 ... ... ... ... ... | 10 |
| Igniter Z.Z.42 ... ... ... ... ... ... | 14 |
| S/60s mechanical time and percussion fuze (Dopp.Z.S/60s) ... | 15 |
| S/60F1. mechanical time and percussion fuze (Dopp.Z.S/60F1) | 19 |
| Tracer from 5 cm. A.P. 40 shot ... ... ... ... | 28 |
| 4·7 cm. Pak(t) and Pak K.36(t) cartridge, Q.F., H.E. ... ... | 28 |
| 4·7 cm. Pak(t) and Pak K.36(t) cartridge, Q.F., A.P.C. ... ... | 29 |
| 7·5 cm. Kw.K., cartridge, Q.F., H.E. ... ... ... ... | 32 |
| 7·5 cm. L.G.40, Q.F. cartridge ... ... ... ... | 34 |
| 7·62 Pak. 36, cartridge, Q.F., H.E. ... ... ... ... | 34 |
| 10·5 cm. L.G.40, Q.F., cartridge ... ... ... ... | 38 |
| 10·5 cm. 1.F.H.18, cast steel H.E. shell (F.H.Gr. 38 Stg). ... | 41 |

### Italian Ammunition

| | |
|---|---|
| Propellant charges ... ... ... ... ... ... | 43 |
| H.E. hand grenade O.T.O. model 35 ... ... ... ... | 44 |
| 25 gram T.N.T. gaine ... ... ... ... ... | 46 |
| 9 grain primer percussion Q.F. cartridge ... ... ... ... | 48 |
| 37/40 cartridge, Q.F., common pointed ... ... ... ... | 48 |
| 37/40 cartridge, Q.F., fuzeless semi-armour piercing ... ... | 51 |
| 65/17 cartridge, Q.F., H.E. ... ... ... ... ... | 53 |
| 70/15 Q.F. separate loading H.E. shell ... ... ... ... | 54 |
| 76/40 cartridge, Q.F., H.E. ... ... ... ... ... | 54 |
| 77/28 Lunga, Corta and 6 Migl. H.E. shell, shrapnel shell and Q.F. cartridges ... ... ... ... ... ... | 58 |
| 105 mm. 28 H.E. shell ... ... ... ... ... | 66 |
| 75 mm. shell. Tabulated details of shell for field and anti-aircraft equipments ... ... ... ... ... | 68 |

### Japanese Ammunition

| | |
|---|---|
| 25 mm. cartridge, Q.F., H.E./T. ... ... ... ... | 70 |
| Direct action nose fuze for Q.F. 25 mm. H.E./T. shell ... ... | 72 |
| 12 grain primer, percussion, Q.F. cartridge ... ... ... | 75 |

# HANDBOOK OF ENEMY AMMUNITION

## GERMAN AMMUNITION MARKINGS AND NOMENCLATURE

The following markings and designations, additional to those given in Pamphlet No. 6, have been met with in the course of examining captured German ammunition :—

### General

The letters " Vz " in the designation of ammunition of Czech origin is the abbreviation for mark or model.

### Fuzes

The letter " S " with an oblique stroke immediately in front of the fuze number was given in Pamphlet No. 6 as an indication of a mechanical time fuze. The same marking has since been found on a time fuze of the combustion type in naval service and thus indicates a time fuze of either type.

The stamping " $Fg^1$ " or " Fl " following an oblique stroke after the fuze number on a mechanical time, or time and percussion, fuze indicates the time mechanism to be of the centrifugally operated Junghans type.

The following letters included in the designation of nose fuzes indicate :—

" l.Igr.Z "     This was incorrectly given in Pamphlet No. 6, as l.I.Gr.Z. May also be found as Jgr.Z.

" Lg.Zdr.S/ "     Time fuze for star shell.

" s.Igr.Z "     Percussion fuze for heavy infantry gun shell.

The abbreviation " Geb " indicates that the fuze is for a mountain equipment.

### Projectile Nomenclature

The dimension included in the designation of spigot mortar bombs is the diameter of the bomb and is not, as in the case of British nomenclature, the diameter of the spigot of the mortar.

The abbreviation " Lg " is used to indicate star shell.

## Projectile Markings

Star shell painted a shade of green (lighter than the deep olive green in common use) with a yellow band above the driving band and a black tip have been met with. This may be a naval shell.

H.E. and smoke mortar bombs may be painted a deep olive green as an alternative to the dull red as given in Pamphlet No. 6.

" Deut " painted on a projectile indicates a coloured smoke round. The colour of the paint used indicates in some cases the colour of the smoke. In other cases, the colour of the smoke is marked on separately, e.g., " blau."

## Bands

The red band above the driving band, referred to in Pamphlet No. 6, page 29, is also found on shell without tracers. The significance of this band is not yet apparent.

## Stencilling

The Arabic numerals stencilled on the head of H.E. and hollow charge shell to indicate the nature of filling are also used on mortar bombs and on piercing shell. On black piercing shell the stencilling is normally in red. The following additional numerals have been met with or reported :—

| Stencilling | Indication |
|---|---|
| 15 | T.N.T./Aluminium 90/10. |
| 17 | T.N.T./Aluminium 90/10 with P.E.T.N./Wax exploders. |
| 33 | P.E.T.N./Wax 85/15. |
| 86 | Ethylene diamine dinitrate. |
| 92 | Cyclonite/Wax 90/10. |

The letters " oM " following the numeral indicate the absence of an exploder container.

The letters " PL " are stencilled on shell of Czech origin to indicate Pentolite.

The following abbreviations are used on labels to indicate the various natures of high explosives :—

| | |
|---|---|
| Fp 02 | T.N.T. |
| Fp 60/40 | Amatol. (The first numeral indicates the percentage of T.N.T.) |
| Gr 88 | Picric Acid. |
| H | Cyclonite. (A numeral after the letter indicates the percentage of wax.) |
| Np | P.E.T.N. (A numeral after the letter indicates the percentage of wax.) |

The letters " F.E.S." stencilled in white above the driving band and, in some instances, on the base of the shell indicate the driving band to be wholly of iron.

## Cartridges

The following additional abbreviations in the designations marked on cases, bags and labels have been found :—

Nz Gew. — Indicating a small arms nitrocellulose powder stabilized with diphenylamine and including ethyl centralite and potassium sulphate.

Np. Gw. — Indicating a small arms, double base propellant of P.E.T.N. and nitrocellulose which includes diphenylamine, ethyl centralite and potassium sulphate.

Nz Man — Indicating a nitrocellulose powder including potassium nitrate and graphite.

Kr.R. — Indicating a central supporting tube of propellant, of comparatively large dimensions, used as a core and support in the arrangement of the propellant charge. The size (length × external diameter × internal diameter) is given in millimetres in the form " (345.14/12) ".

N.P. — Indicating solid cylindrical granular propellant. The size (length × diameter) in millimetres is given in the form " (1,5 · 1,5) ".

Pl.P. — Indicating propellant in the form of discs of comparatively large diameter with multi perforations. The size (diameter × thickness) in millimetres is given in the form " (50 · 0,2) ".

Stb.P. — Indicating propellant in the form of small cylindrical grains. The size (length × diameter) in millimetres is given in the form " (0,8 · 0,8) ".

Code letters for the identification of the shape of Digl. propellant in sectional charges of the howitzer type are met with on charge bags, labels and packages. The letters used and their significance are as follows :—

D — Digl. propellant in the form of flakes or perforated discs (washer form).

DR — Digl. propellant in tubular form.

DST — Digl. propellant in strip form.

These letters are found adjacent to the number of the charge section on the bags (*see* Pamphlet No. 6, Fig. 17), in bold lettering on labels and stencilled in white on the package, usually on the lid.

The letters " o.B.D.", stencilled in black on the side of the case of a fixed Q.F. round are used to indicate the absence of a decoppering agent from rounds in which this component is normally used.

## Standard Charge Temperature

The red stencilling " Abgebr.Ldg." on the side of a cartridge case (referred to in Pamphlet No. 6, page 35, sub-para. (vii) ) indicates that the weight of the propellant charge has been reduced for use in hot climates. The standard temperature on which the reduction is based is indicated by the stencilling " Schusstafelm P.T. + 50° C.", also in red. This standard was superseded by that of 25° C.

## GERMAN S-MINE 35
### (S.Mi. 35)
### (Fig. 1)

The mine, containing steel balls and a bursting charge of H.E., is operated by a contact pressure fuze or by pull-igniters with trip-wires and is projected about 4 feet from the ground where it functions. The radius of the area over which the bullets are dispersed extends up to about 200 yards. The igniter S.Mi.Z.35 is described in this pamphlet. The use of an electrical igniter, E.S.Mi.Z.40, has also been reported.

The cylindrical exterior, painted the normal deep olive green, is 4 inches in diameter and 5·2 inches high with a closing plate at the top. The plate has a protruding screwed cap of plastic at the centre and carries three small equally spaced screw plugs and one larger screw plug. The overall height is 6 inches and the filled weight, 9 lb. 7 oz.

The mine consists of the following main parts :—

(a) The mortar or outer casing.

(b) The body which is a sliding fit in the mortar and contains the steel balls and an H.E. bursting charge.

(c) The ejection charge contained in a recess in the base of the body.

The mortar is in the form of a cylinder, open at the top, and is pressed from mild steel of 0·079 inch thickness. A light metal ring which overlaps the top plate is soldered around the exterior at the mouth.

The body consists of an outer and an inner seamless cylinder of mild steel which are held at the top and bottom between the machined circular cast iron top plate and base plate. The plates are secured by a central tube of mild steel which passes through their centres and is fitted with a nut which bears on the upper surface of the top plate and a flanged adapter which screws into the tube at its lower end and bears against the lower plate. The top plate is 0·39 inch thick and has what appears to be a filling hole for a cast bursting charge and three holes for the insertion of the detonators. The three holes are closed by screwed plugs, the plug for the filling hole being larger than the others. Three brass tubes, forming detonator pockets, are

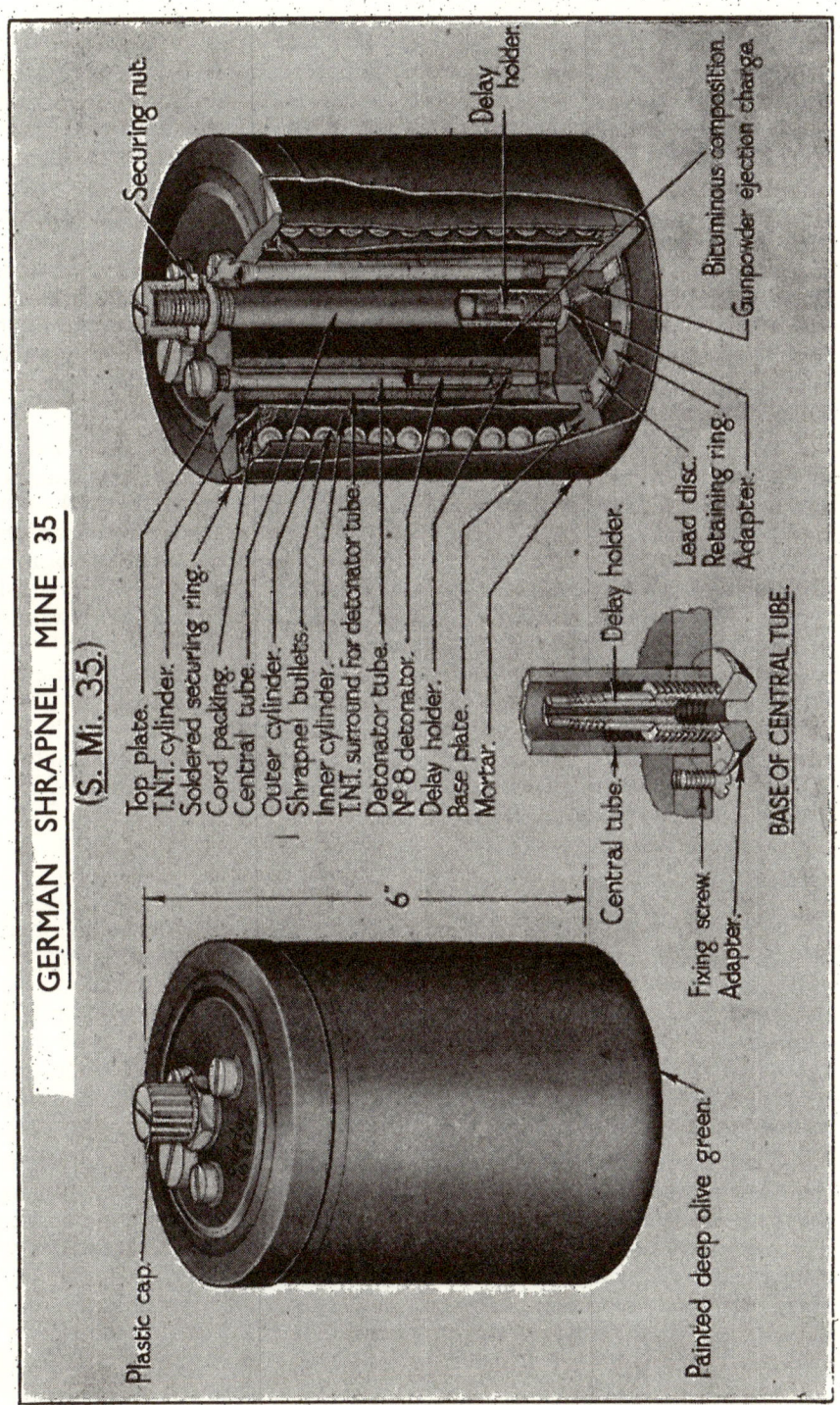

Fig. 1

fitted between the holes in the top plate and delay holders screwed into similar holes in the base plate. These holes lead into a circular recess in the underside of the base plate which contains the ejection charge. The adapter at the lower end of the central tube also leads into this recess at the centre and has a delay holder screwed into it from the top. The recess is closed at the base by a lead disc which may be flat or concave. The disc is held by a retaining ring of mild steel secured by screws.

The steel balls contained between the inner and outer cylinders number approximately 350 and are of high carbon steel containing some chromium and nickel. The balls are about 0·35 inch in diameter and weigh about 129 to the pound. The V.D. Hardness figure at the surface averages 830. A packing of cordage is inserted between the top of the balls and the top plate.

The T.N.T. bursting charge weighs approximately 7 oz. 13 drs. and is in the form of a block filling consisting of a cylinder of suitable size for insertion in the body. The wall of the cylinder is 0·24 inch thick and the inside contains a core of bituminous composition in which there are three smaller cylinders of T.N.T. to form surrounds for the detonator tubes also an axial channel to fit over the central tube. The exterior of each T.N.T. cylinder is waxed and the whole is contained in a waxed paper carton. The total weight is 1 lb. 0 oz. 4 dr. The delay holders are of brass and contain a pressed filling of delay composition. The ejection charge in the recessed base plate consists of approximately 45 grains of gunpowder. The weight of the body (excluding the mortar) filled and fuzed is approximately 7 lb. 13 oz. Oil is used as a lubricant between the interior of the mortar and the exterior of the body.

## Action

The mine is prepared for use by inserting a No. 8 detonator, open end downwards, into each of the detonator tubes, removing the plastic cap and substituting the fuze S.Mi.Z.35. When trip action is required, a Y-shaped adapter is screwed on the top of the central tube instead of the fuze and pull-igniters of the Z.Z.35 type (*see* Pamphlet No. 8) are inserted in the arms of the adapter and connected to the trip-wires. When the electrical fuze is used, a number of mines can be fired simultaneously from a distance.

When the fuze or a pull-igniter, as the case may be, is functioned, the flash passes down the central tube and ignites the composition in the central delay holder which in turn ignites the ejection charge. The explosion of the charge ignites the composition in the other delay holders and ejects the body from the mortar. Whilst in flight the delay composition burns through the length of the holders and initiates the detonators with the result that the bursting charge is detonated and the balls projected.

## Modifications

The discovery of an S-Mine 35 with the following modifications has been reported:—
   (a) Detonator tubes of cardboard instead of brass.
   (b) Shrapnel, consisting of short lengths of ⅜ inch mild steel rod instead of hardened steel balls.

Mines may be found to have three screws in the base of the mortar which are inserted from the exterior to engage with corresponding tapped holes in the base plate of the body. At a later date the use of the screws was discontinued, and the holes in the base of the mortar were closed by soldering. The base plates with the tapped holes continued to be used.

### GERMAN S-MINE IGNITER 35
### (S.Mi.Z. 35)

(Fig. 2)

The igniter, mainly of aluminium, is brown in colour as the result of varnishing and is cylindrical with three steel prongs in the top of the plunger which protrudes at the head. For safety in transport a lateral safety bolt is fitted in the plunger where it emerges from the body.

The aluminium body is 0·7 inch deep in diameter and is made up of three sections screwed one into the other. The top section forms a casing for the supporting spring and has a hole in the head for the plunger. At its lower end it is screwthreaded internally to receive the middle section. After assembly the two sections are locked by a plug inserted between coinciding semi-circular recesses formed in the two sections. The middle section has a band of milling around the centre of its exterior. Internally, at the upper part, it is reduced in diameter to locate the two steel retaining balls in the plunger and to act as guide for the plunger. Screwthreads are formed externally at the upper end for insertion in the top section and internally at the lower end to receive the bottom section. The bottom section also has a milled band and is locked to the middle section by a plug inserted between semi-circular recesses. A diaphragm in the top of this section has a central hole in which the cap assembly is fitted. The lower end is screwthreaded internally for assembly with the mine. For transport the fuze is closed at the base by a screwed plug of the moulded plastic type.

The aluminium plunger is cylindrical with an external flange at the centre which limits its upward movement and engages one end of the supporting spring. From the base the plunger is recessed to accommodate the striker with its spring and has two radial holes in which steel balls are located to engage a groove in the striker. In the upper part of the plunger there is a lateral channel to receive the safety bolt. A screwthreaded channel, drilled from the head, leads into the centre of this channel and contains a spring-loaded ball which engages a groove in the centre of the bolt. The ball and spring are retained in position by a screwed plug which also retains two steel washers securing the steel prongs. The prongs are inserted in holes in the head of the plunger and are splayed by notches in the inner circumference of the lower washer and in the outer circumference of the upper washer.

The safety bolt is of aluminium and has a steel split ring at one end and a retaining nut at the other. A groove of semi-circular section is formed in the bolt to engage the ball in the head of the plunger.

The steel spiral supporting spring fits around the plunger below the flange and is held in compression between the flange and the upper end of the middle section of the body.

The steel striker has a stem formed on its upper end over which its steel spiral spring is assembled. The spring is held under compression between the body of the striker and the end of the recess in the plunger. A groove, with the upper side inclined, is formed in the striker body to engage with the two steel retaining balls.

The cap assembly consists of a copper sheath containing the cap and the anvil. The sheath is flanged at the top and is turned in at the base to secure the anvil. The anvil is formed by a bent brass strip with half collars formed on each end under which the sheath is turned inwards. The copper cap contains 0·8 grains of composition consisting of 40·8 per cent. of lead styphnate, 40·7 per cent. of barium nitrate, 2·9 per cent. of antimony sulphide and 15·6 per cent. of calcium silicide.

**Action**

The safety bolt is withdrawn by removing its retaining nut and pulling the bolt out of the plunger against the resistance of the spring-loaded ball.

When pressure is applied to the prongs the plunger is forced down into the body against the resistance of the supporting spring. When the two steel balls in the plunger have moved down sufficiently to lose the support of the low diameter portion of the body, these are ejected by the inclined surface in the striker groove, under the pressure of the striker spring, which drives the striker on to the cap assembly. The weight required to operate the fuze is approximately 7 lb.

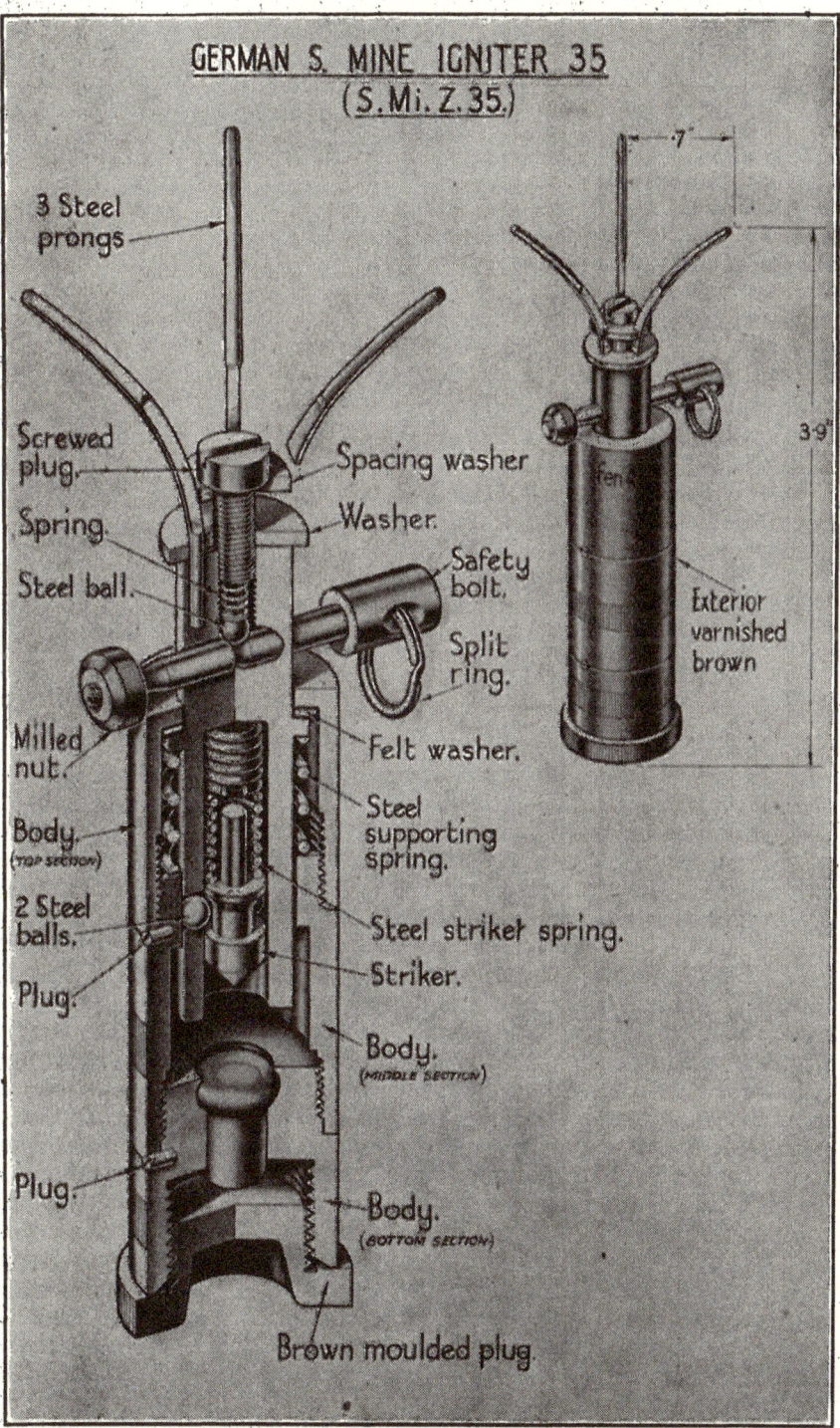

Fig. 2

## WOODEN BOX MINE 42
### German designation : Holzmine 42
(Fig. 3)

### Data

| | |
|---|---|
| Internal dimensions ... | 290 × 272 × 82 mm. (11·4 × 10·7 × 3·2 in.) |
| Size of lid ... ... | 13 × 12 × ¾ in. |
| Size of aperture in lid... | 163 × 65 mm. |
| Pressure block (without feet) ... ... | 6 × 2·4 × 2·4 in. |
| Size of compartments : | |
| Main charges ... | 116 × 196 mm. |
| Priming charges ... | 45 × 175 mm. |
| Thickness of partitions... | 5 mm. |
| Size of shearing flange ... | 195 × 32 × 14 mm. |
| Size of slot in shearing flange ... ... | 7 × 17 mm. long. |
| Size of wooden blocks ... | 79 × 20 × 22 mm. deep. |
| Main explosive ... ... | Amatol 50/50. |
| Total weight of explosive | 11·9 lb. |
| Total weight of mine ... | 18 lb. |

*Note :* Where appropriate the first dimension is parallel to the length of the pressure block.

### Casing

The mine is contained in a box (1) of ¾ inch timber to which the base is nailed. The box is sub-divided internally into four compartments, of which the two at the side contain the main explosive filling (2), while the central compartment (3) contains the primer charges (3-200 gm. Sprengkörper) and the end compartment (4) the operating mechanism. The central compartment contains a bottom packing piece 34 × 10 mm. thick, two small wooden blocks, one, 34 × 20 × 8 mm. (5), nailed to the bottom packing piece and the other, 34 × 22 mm. section (6), nailed to the back ; these locate the primer charges firmly in place. The partitions (7) and the partition (13) are removable. The end compartment (4) contains a shearing flange (8) secured to the outside wall by two ⅜ inch wooden dowels ; this flange is provided with a central slot (9) which allows the end of the striker to pass when arming. Two wooden blocks (10) are nailed to the base on each side of the igniter rest and carry the pressure block when the mine is not armed. The igniter rest (12) consists of a small piece of wood with a U-shaped piece cut-out at the top and screwed to the base from the underside by two large screws so that the " U " is opposite the slot which is cut in the partition (13). The lid (14) is secured to the mine at the front and back by metal hooks (15), and is located by wooden dowels (16). At one end is a rectangular hole (17) for the pressure block (11), and at the other end a stacking piece (18). On the underside of the lid are two 22 × 9 mm. diameter, wooden

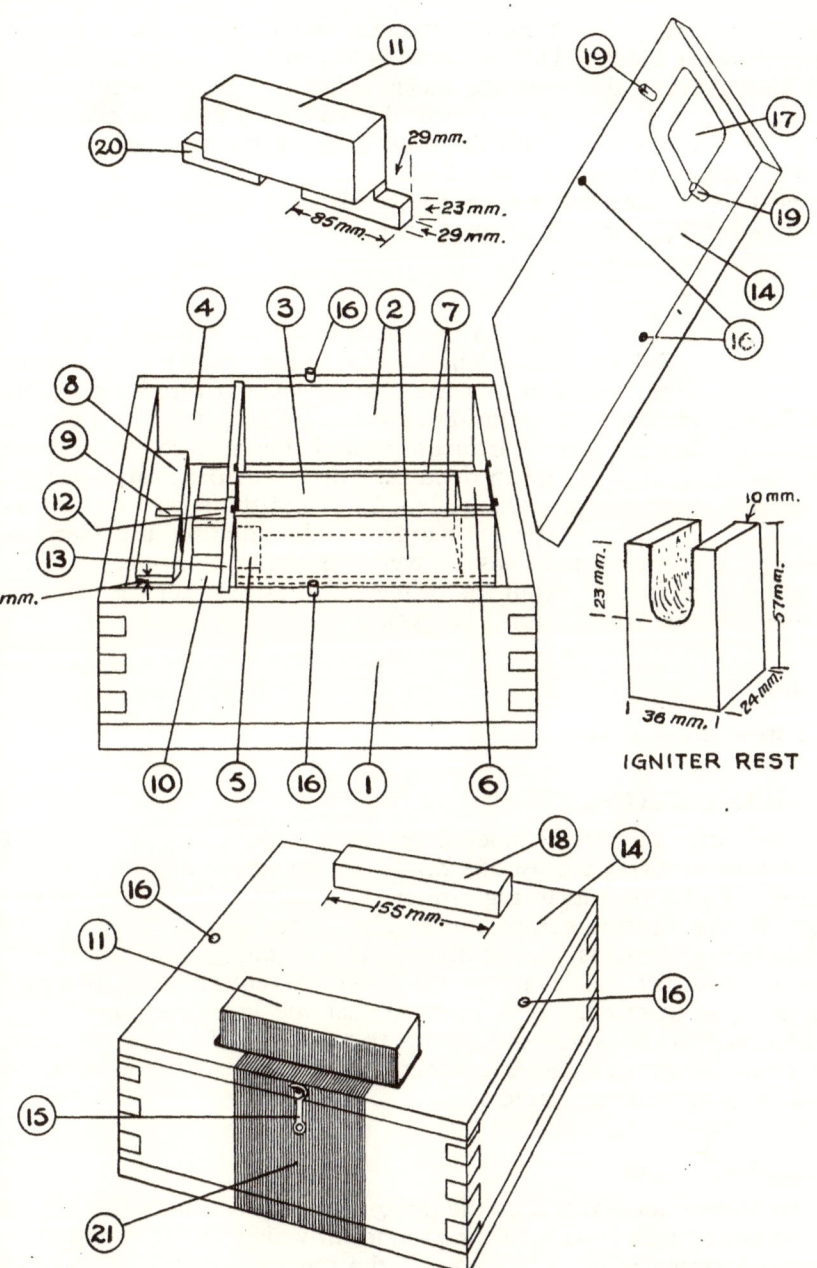

# GERMAN WOODEN BOX MINE 42

Fig. 3

pegs (19) which prevent the two wooden feet (20), on the underside of the pressure block (11), from moving towards the centre of the box and also prevent the pressure block being put in the wrong way round. When the mine is armed the feet (20) rest on the shear flange (8), in which position the head of the pressure block projects about 2 inches above the lid. During transit the pressure block is reversed so that the feet rest on the blocks (10); in this position the pressure block protrudes about ¾ inch.

### Filling

The main filling consists of two charges of Amatol 50/50, each of approximately 2·4 kg. (5·3 lb.) in the compartments (2). Specimens captured in Sicily revealed that the charges were covered with some bitumastic substance for exclusion of water. The primer charges are situated in the central compartment (3), and consist normally of three 200 gm. charges. An Italian document describes the charge nearest the igniter as a " 200 gm. charge No. 28 ", and the remaining two as " Standard 200 gm. prepared charges ". In addition the German marking (*see* below) mentions " Zwei Füllkörper o. Bohrung". In the absence of further detail the primer charges have all been shown as Sprengkörper 28 (200 gm. charges).

### Igniter

This is the Z.Z.42.

### Anti-lifting Devices

These can be fitted underneath the central compartment and would most probably consist of a Z.Z.35 igniter screwed into a 200 gm. charge. To fit an anti-lifting igniter the charges require rearrangement in the central compartment (3). The charge carrying the Z.Z.42 igniter remains, one additional 200 gm. charge is laid with the hole downwards; the length of the latter prevents a third 200 gm. charge being inserted. It is believed that the holes for anti-lifting igniters are made when the booby traps are laid. Reports from Sicily say that of the mines encountered in one area, 35 per cent. were fitted with anti-lifting igniters.

### Method of Laying

An Italian document says the mine is usually laid with the top of the pressure block level with the ground, in which case the minimum spacing between mines is 6 feet. If the mine has to be laid on the surface this spacing must be increased to 13 feet. The mine is laid with the red strip (21) facing defending troops, i.e. so that the H.E. charge lies more directly under the oncoming load. The mine cannot be laid in water or in damp ground, where it would deteriorate.

## Method of Operation and Laying

Pressure on the pressure block (11) shears the ⅜ inch dowels securing the shear flange (8) to the outer wall and forces the shear flange down on to the igniter pin, which is withdrawn and frees the spring-loaded striker.

## Identification

A vertical red band (21) is painted down the centre of one end of the box and is continued on to the lid. When armed, the side of the pressure block (11) facing this band is also painted red.

The mine is marked as follows :—

I

HOLZMINE 42

bestehend aus

Zwei Ladungen zu je etwa 2·4 kg. Fp. 50/50

Zwei Füllkörper o. Bohrung

einem Sprengkörper 28 m. Bohrung

gue

eingehörige Zündungen in besonderem Packgefäss

## Translation

I

WOODEN MINE 1942 PATTERN

consisting of

two charges each of approximately 2·4 kg. 50/50 Amatol

two " filling " charges without igniter socket,

one 1928 pattern 200 gm. charge with igniter socket

gue (manufacturers code)

relevant firing assembly in a special packing container.

## Firing Pressure

It is reported from Sicily that the mine will fire under a pressure of 200 lb., whereas the firing pressure given in the Italian document is 200-300 kg. (440-660 lb.). From the diameter of the shear dowels it appears that the figure of 200 lb. is the more probable.

## To Neutralize

(i) Look for and neutralize any anti-handling devices.
(ii) Remove lid carefully.
(iii) Reverse pressure block so that it sits in the safe position.
(iv) Replace lid.

## To Disarm

(i) Look for and neutralize any anti-handling devices.
(ii) Remove lid carefully.
(iii) Remove pressure block.
(iv) Hold the actuating pin in position and remove one of the blocks of the priming charge but not the one into which the igniter is screwed. Slide back the block with the igniter until the actuating pin is clear of the shearing flange—lift out charge and igniter.
(v) Carefully unscrew the igniter, holding the pin in position.
(vi) Remove detonator.

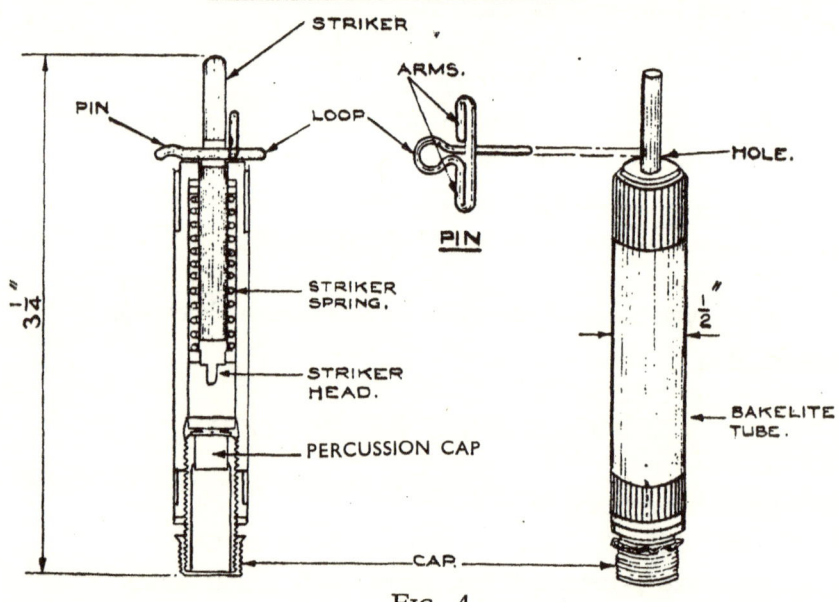

Fig. 4

## Detectability

The mine is detected with difficulty by the U.S. detector at a distance of 6-8 inches under normal conditions. Although the igniter body is bakelite the striker and pin are metal and a specimen of the mine when examined was found to contain 32 nails of varying sizes, 2 screws, and 2 fastening clips.

### GERMAN PULL IGNITER 42
### (Z.Z.42)
(Fig. 4)

This igniter is used in the wooden box mine described elsewhere in this pamphlet and also in the Stock-mine and Schü-mine 42 and wooden box mine U.B.1.

## Construction

The igniter consists of a bakelite tube about 2.5 inches long, 0.5 inch in diameter carrying a spring-loaded striker. The tension of the spring retains a cross shaped pin in a hole in the striker rod, firmly against the top of the bakelite case. The percussion cap is carried in a tube screwed into the base of the igniter. The tube is closed by a screwed cap. No safety device is fitted. The screwed cap is removed and a detonator inserted when required for use.

## Operation

Pressure on the arms of the pin, or tension applied to the loop of the pin will cause it to withdraw. The force required is reported to be between 6 and 10 lb. weight. The striker is then impelled forward on to the percussion cap by the action of the spring.

## GERMAN MECHANICAL TIME AND PERCUSSION FUZE S/60s
### (Dopp.Z.S/60s)
### (Figs. 5, 6 and 7)

The fuze which has already been described briefly in Pamphlet No. 1 is used in H.E. shell for the following equipments :—

   Geb.G.36 (7·5 cm. mountain gun)
   F.K. 16n.A. (7.5 cm. field gun) (obsolete).
   l.F.H. 16 and 18 (10·5 cm. howitzers)
   10 cm. K.17 and K.17/04 n.A. (10.5 cm. field guns) (obsolete).
   s 10 cm. K.18 (10.5 cm. medium gun)
   lg.s.F.H.13 (15 cm. howitzer) (obsolete).
   s.F.H.18 (15 cm. howitzer).

The external appearance of the fuze is the same as that of the Zt.Z.S/30, described in Pamphlet No. 8, except that it bears the stamping " Dopp Z S/60s " just above the flange. The time mechanism of the fuze is of the clock spring type and is the same as that described for the Zt.Z.S/30 but has a time of running of 60 seconds.

The percussion mechanism is of the graze type and is contained in a central recess below the time mechanism. The recess has a flash hole at the base and has two increases in diameter towards the mouth. The mouth is screwthreaded to receive the needle disc.

The graze mechanism is of the type commonly used in German nose fuzes and consists of a brass inertia pellet, with detonator, which is held off a fixed needle by a creep spring and four centrifugal segments. The segments are pivoted on pins in the platform formed by the change in diameter, and are encircled by the usual expanding spring ring.

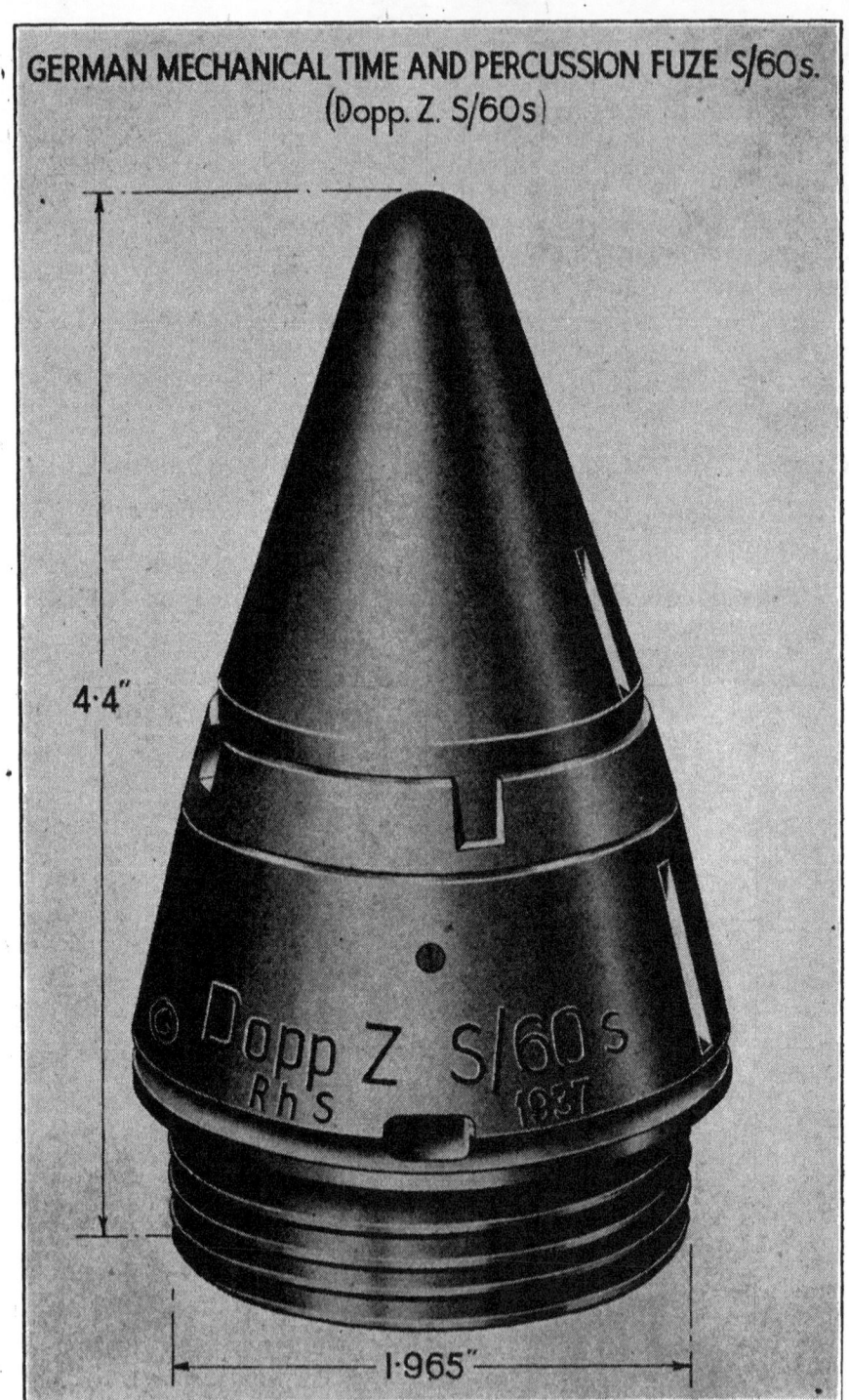

FIG. 5

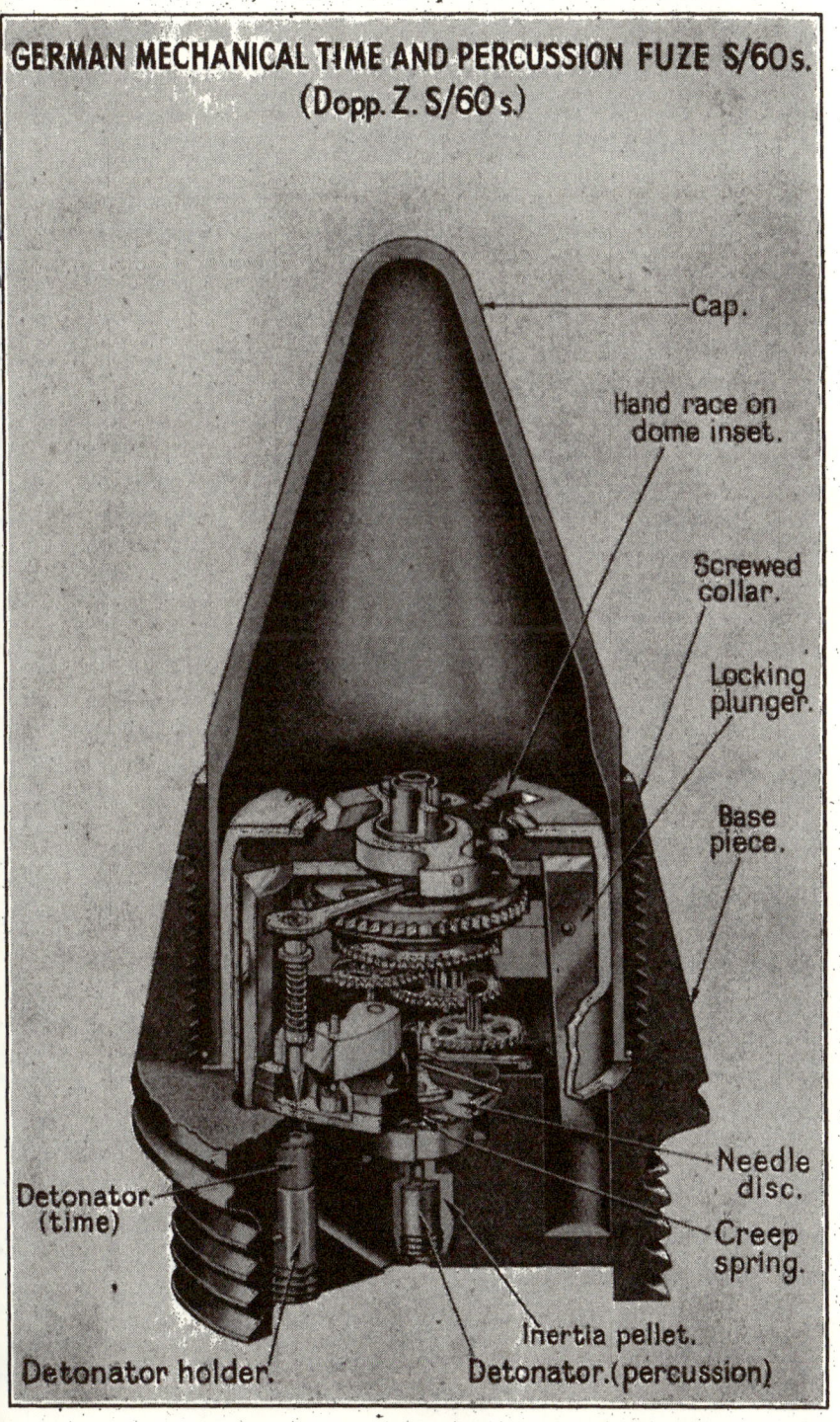

Fig. 6

# GERMAN MECHANICAL TIME AND PERCUSSION FUZE S/60s.
## Dopp. Z. S/60s.
## DIAGRAMMATIC ARRANGEMENT OF MECHANISM

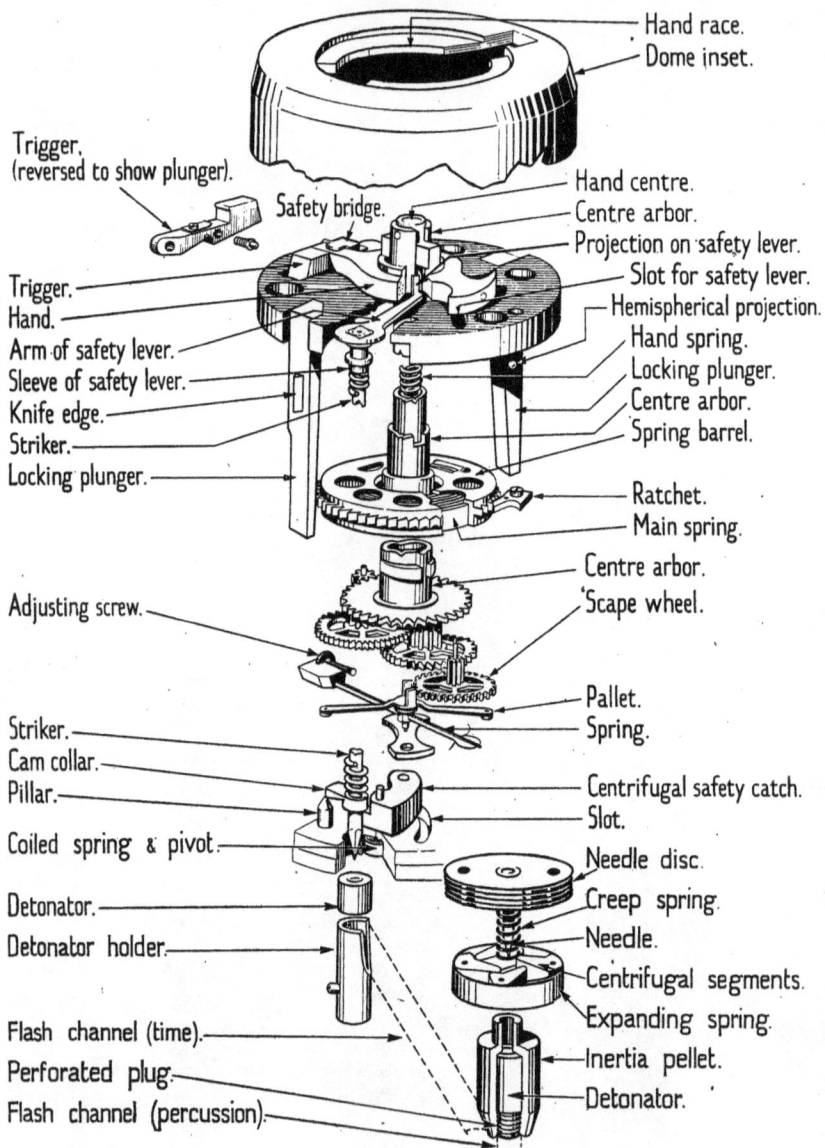

Fig. 7.

## Action

The time action is the same as that described for the Zt.Z.S/30 in Pamphlet No. 8.

The fuze is set to the safety mark on the graduated setting key when percussion action is required. When set at this position, the cut away part of the hand race at the top of the dome inset is masked by the safety bridge. Thus, although the hand is released by the trigger when acceleration occurs and is rotated by the mechanism during flight, it cannot rise to release the striker. The percussion arrangement is armed by centrifugal force during flight by the enlargement of the expanding spring ring which permits the pivoted segments to be swung clear of the inertia pellet. The pellet is prevented from creeping during the period of deceleration by the creep spring between the needle disc and the pellet. On graze the momentum of the pellet overcomes the spring and the detonator is impinged on the needle.

## GERMAN MECHANICAL TIME AND PERCUSSION FUZE S/60 Fl
### (Dopp. Z. S/60 Fl)
(Figs. 8, 9 and 10)

The fuze is used in H.E. shell and the F.H.Gr.40 Nb (smoke shell) for the 10·5 cm. 1.F.H.18 (gun-howitzer).

The time mechanism is of the centrifugally operated type with a maximum time of running of 60 seconds and is designed to prevent functioning at settings of less than approximately one second. The explosive content consists only of an igniferous detonator which serves for both the time and the percussion action. When functioned by percussion, the pellet carrying the detonator is released from the control of the time mechanism on impact.

The external appearance of the fuze differs from the mechanical time fuzes of the S/30 type (*see* Pamphlet No. 8) in that the top of the cap is flat and is closed by a metal disc. The fuze is identified by the stamping " Dopp.Z.S/60 Fl.", just above the flange. The stamping " Fl " and the flat top distinguishes this centrifugally operated type from the Dopp.Z.S/60s which is similar in appearance but is operated by a clock spring.

The weight of the fuze is approximately 1 lb.

The main components of the fuze consist of the body or base piece, the cap base, the cap, the time mechanism and the percussion mechanism.

The base piece is screwthreaded externally below the flange for insertion in the shell and has a large cylindrical recess in which the time mechanism is located. A brass cylinder is fitted to line the wall of the recess and below this cylinder a circular groove is formed in the wall to correspond with a similar external groove in the cap base. A

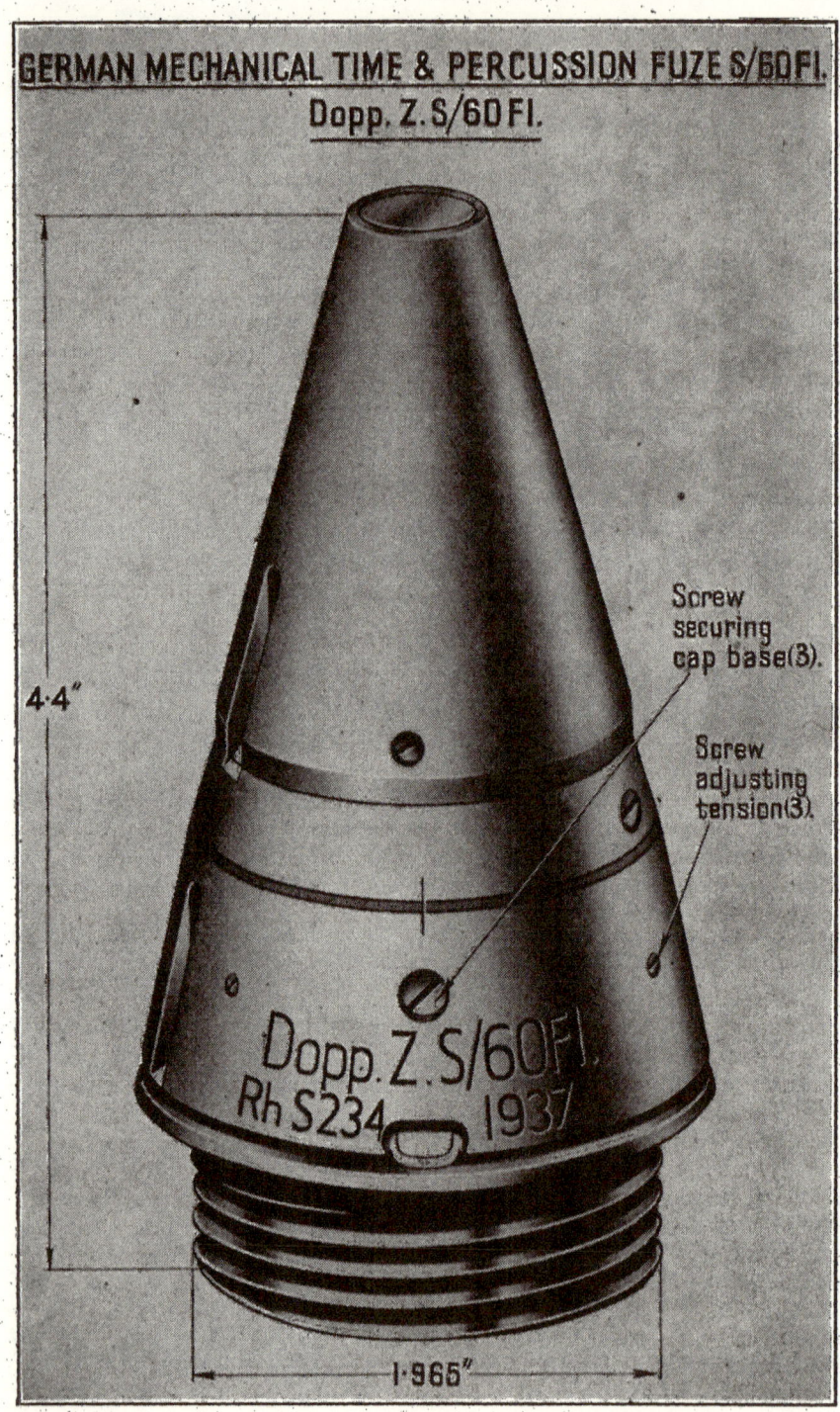

Fig. 8

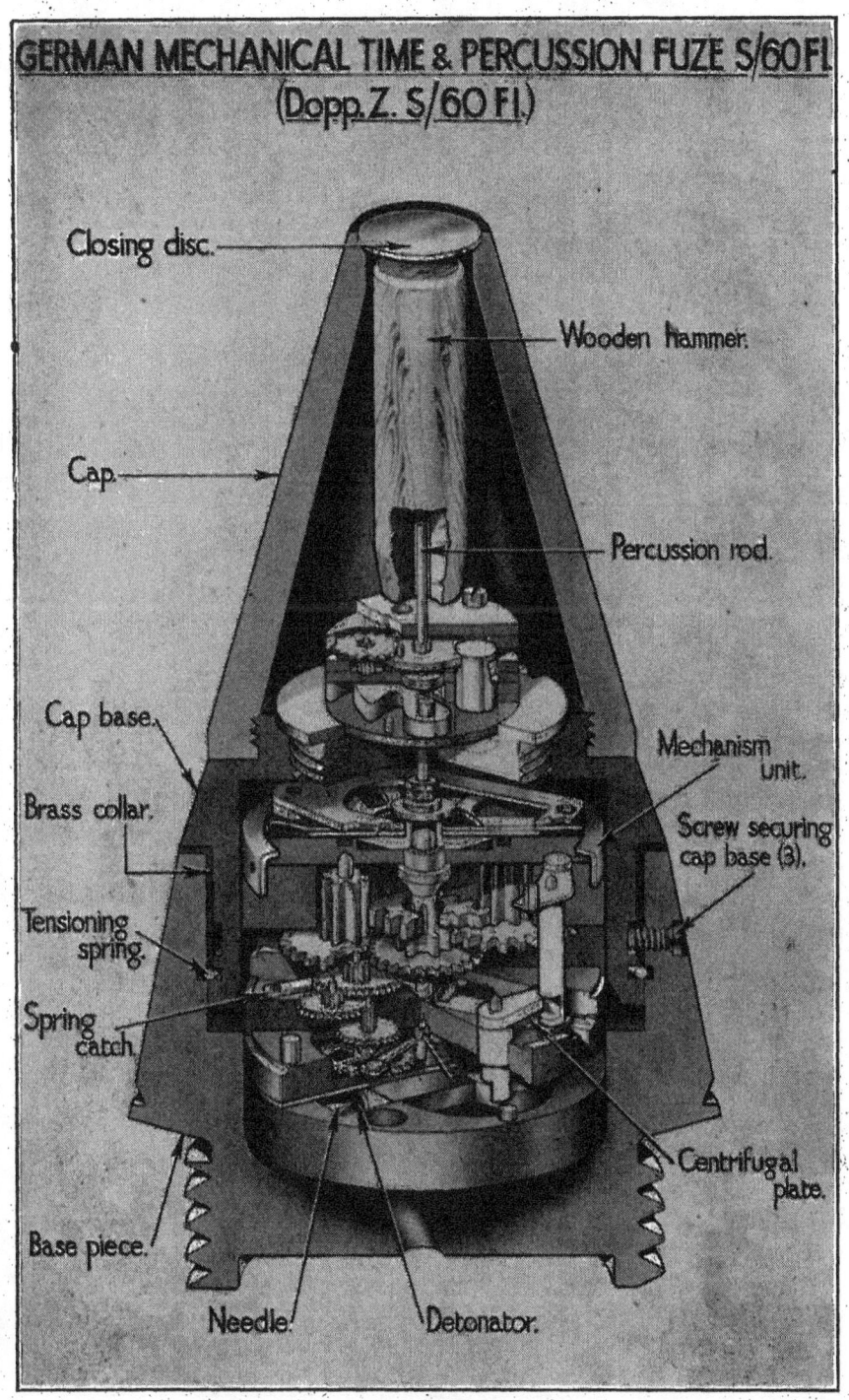

Fig. 9

# GERMAN MECHANICAL TIME & PERCUSSION FUZE S/60 Fl.
## (Dopp Z. S/60 Fl.)
### DIAGRAMMATIC ARRANGEMENT OF MECHANISM.

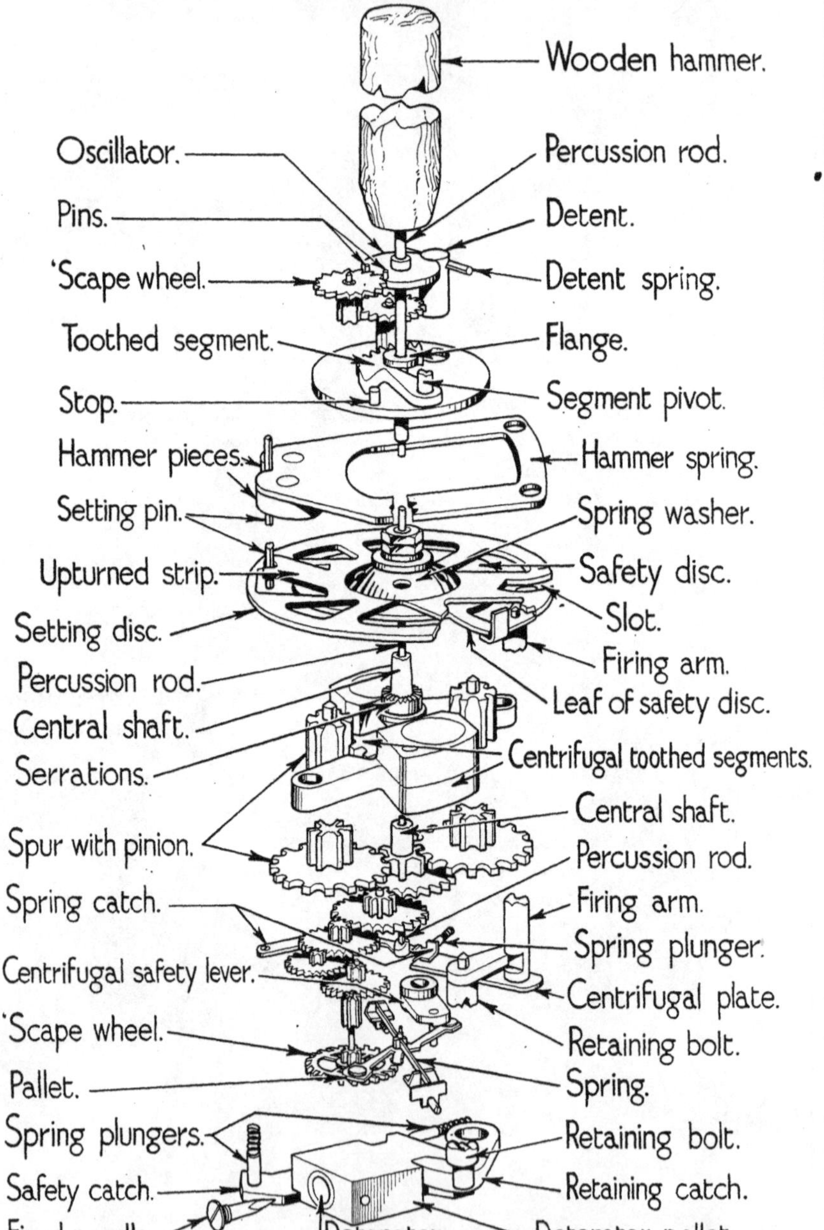

- Wooden hammer.
- Oscillator.
- Percussion rod.
- Pins.
- Detent.
- 'Scape wheel.
- Detent spring.
- Toothed segment.
- Flange.
- Stop.
- Segment pivot.
- Hammer pieces.
- Hammer spring.
- Setting pin.
- Spring washer.
- Upturned strip.
- Safety disc.
- Setting disc.
- Slot.
- Percussion rod.
- Firing arm.
- Central shaft.
- Leaf of safety disc.
- Serrations.
- Centrifugal toothed segments.
- Spur with pinion.
- Central shaft.
- Spring catch.
- Percussion rod.
- Centrifugal safety lever.
- Firing arm.
- 'Scape wheel.
- Spring plunger.
- Pallet.
- Centrifugal plate.
- Spring plungers.
- Retaining bolt.
- Safety catch.
- Spring.
- Fixed needle.
- Retaining bolt.
- Detonator.
- Retaining catch.
- Detonator pellet.

FIG. 10

spring wire with four semi-circular bends is fitted within these two grooves to tension the cap base and the cap.  For the adjustment of the tensioning during assembly, three small screws are fitted from the exterior at equally spaced positions around the groove in the body. Semi-circular recesses are formed in the groove at the points where the screws bear on the bends in the tensioning spring.  Three larger screws are also inserted through the wall of the body and the brass liner to engage a second groove in the cap base and thus secure it to the body without preventing the rotation necessary in setting the fuze. Inside the recess, at the base, there are two small recesses to receive corresponding locating studs on the base of the time mechanism unit. An inclined flash channel is formed near the centre of the base and three holes are provided for the bolts securing the time mechanism unit.

The cap base consists of an aluminium sleeve which fits into the recess in the body and surrounds the time mechanism unit.  The sleeve is flanged near the top to coincide with the contour of the fuze. Above the flange it is reduced in diameter and screwthreaded externally to receive the cap and internally to receive the percussion mechanism unit.  The internal shoulder formed by the reduction in diameter carries the setting pin and the hammer spring.  Two circumferential grooves are formed on the outside of the sleeve near the base. The lower groove receives the wire tensioning spring, the upper groove is engaged by the securing screws in the body.

The cap is cone-shaped with a flat top which is closed by a metal disc.  Internally, at the base, it is screwthreaded for assembly with the cap base.

The time mechanism is assembled in a cylindrical unit of superimposed plates of brass and aluminium and consists of a central shaft rotated by two centrifugally operated segments and controlled by an escapement.  The shaft carries a slotted setting disc which releases the firing arm when the slot in the disc has been rotated into alignment with the arm.

The central shaft is tubular with a spur near the lower end and a pinion at the base.  The upper part of the shaft is reduced in diameter to receive the bush carrying the setting and safety discs and is screwthreaded to receive the tensioning and locking nuts.  The sloping shoulder formed by the reduction in diameter is serrated to engage with similar serrations on the bush so that the bush is locked to the shaft.  The percussion rod of the percussion mechanism passes through the tubular shaft.

Two weighted centrifugal toothed segments, each pivoted on a semi-circular disc, are enmeshed with spurs which, through pinions formed at their base ends, drive the central shaft.

A train of four spurs and pinions transmits the motion of the pinion at the base of the central shaft to the escapement wheel.  The wheel is engaged by two vertical arms on the pallet which is weighted at each end and controlled by a straight adjustable hair spring.  The pallet is locked at one end by a centrifugal safety lever which is fitted with a weighting pin and held by a retaining spring. A step formed on the pivoted end of the lever is engaged by the end of the spring when the lever has swung to the armed position.

The setting disc, with the safety disc beneath it is carried by a bush which fits over the upper end of the central shaft. The bush is in the form of a sleeve with a hemispherical flange at the base. The sleeve is serrated at the lower end to engage a corresponding shoulder on the shaft so that it rotates with the shaft. The safety disc is keyed to the flange of the bush and is of smaller diameter than the setting disc but has a projecting leaf at one part of its circumference which can close the slot in the setting disc and so prevent the operation of the firing arm. The setting disc is capable of rotation on the bush but is tensioned by a saucer-shaped spring washer which is compressed by the tensioning and locking nuts. An upturned forked strip on the setting disc engages the setting pin in the cap base above. A curved slot, diametrically opposite the strip, provides clearance for a projection on the firing arm and thus permits the arm to rotate when these two are in alignment.

The firing arm is positioned near the periphery of the mechanism unit and consists of a short vertical shaft with a small pivot at each end. The upper end of the shaft is fitted with a crosshead which is weighted at one end and has a vertical projection at the other. Near the base, the shaft is cut away to provide clearance for the arm of the retaining bolt when the shaft is turned. The projecting pivot at the base of the shaft is held by a centrifugal plate. The plate is held at its inner end by a spring which engages a vertical projection on the plate. The spring catch consists of a bent strip secured at its outer end and shaped at the inner end to engage the centrifugal plate. A ramp, or inclined projection, near its inner end is in contact with the base of the percussion rod. A spring plunger bears on the catch where it engages the plate and assists in keeping the catch engaged.

The retaining bolt consists of a short vertical shaft with an arm at the top which is held by the firing lever. Near the base of the shaft a flat surface is formed which keeps the retaining catch in engagement with the detonator pellet.

The retaining catch consists of a flat hook which is pivoted at one end and engages a notch in the detonator pellet at the other.

The detonator pellet is contained opposite to a fixed needle in a transverse rectangular slot in the base of the mechanism unit and consists of a rectangular brass pellet carrying a detonator at one end. Near the other end, which is held against a spring plunger, two vertical notches are cut in the sides. A flash channel leads from the detonator to the base of the pellet. In addition to the retaining catch, the pellet is also held by a centrifugal safety catch.

The safety catch, located by a spring plunger in a channel cut in the side of the pellet slot, consists of a flat brass plate shaped at its inner end to engage a corner of the detonator pellet. A recess is formed in the catch to receive the rounded head of the spring plunger.

The percussion mechanism is contained mainly in the cap and consists of a wooden hammer fitted over a percussion rod which is supported by a centrifugally operated mechanism under the control of an escapement. The lower end of the rod bears against the inclined projection on the spring catch and when pushed in releases the centrifugal plate. The outward movement of the plate results in the release of the detonator pellet.

The percussion rod fits into a recess in the base of the wooden hammer and has a flanged sleeve fixed near its upper end. The flange is supported by a toothed segment which is pivoted at one end and is enmeshed with a spur carrying a pinion. The pinion is enmeshed with a spur carrying the escapement wheel which works in conjunction with two pins on an oscillating disc. The disc is fitted loosely over the upper end of the percussion rod and has two semi-circular recesses in its periphery. One of these recesses has a pin near each end to engage the escapement wheel, the other receives one side of detent in the form of a solid cylindrical pellet with a notch in its outer side. The detent is supported by a straight wire engaging the notch.

## Time Action

The time of running is governed by the size of the arc extending clockwise between the curved slot in the setting disc and the position of the vertical projection on the crosshead of the firing arm. The fuze is set by turning the cap with the aid of a graduated setting key. The rotation of the cap with its base is transmitted to the setting disc by the setting pin engaging in the forked strip of the setting disc whilst the safety disc is held keyed to the stationary bush on the central shaft. The curved slot in the setting disc is thus rotated clear of the projecting leaf on the safety disc. The width of the leaf and its position relative to the projection on the crosshead of the firing arm are so arranged that the leaf still closes the slot at settings up to approximately one second and so prevents the fuze functioning dangerously near the gun.

On acceleration the hammer spring sets back and flattens down the forked strip on the setting disc, thus disengaging it from the setting pin. During flight the centrifugal safety lever is swung clear of the pallet and releases the escapement mechanism. The toothed segments, operated by centrifugal force, then swing outwards and, through their spurs and pinions, rotate the central shaft with the disc assembly under the control of the escapement. Whilst this movement is in progress the weighted end of the crosshead on the firing arm tends to swing outwards but is prevented from so doing by the vertical projection at the other end bearing against the edge of the rotating setting disc. Also, the spring plunger of the safety catch is eased from the catch by deceleration and the catch is thrown clear of the detonator pellet by centrifugal force.

When the slot in the setting disc has rotated into alignment with the vertical projection on the crosshead the firing arm is revolved to the extent permitted by the slot and the recess cut in its lower part is turned to provide clearance for the arm of the retaining bolt. The arm then swings outwards and the flat surface on the shaft releases the retaining catch. The catch is thrown clear of the detonator pellet which is driven on the needle by its spring plunger.

**Percussion Action**

With the fuze set for percussion, that is to the safety mark on the graduated setting key, the leaf on the safety disc closes the slot in the setting disc. Thus although the disc assembly is rotated during flight, the two discs move together and the slot remains closed.

On acceleration the detent locking the oscillator sets back and releases the oscillator. The toothed segment supporting the flange on the percussion rod is swung outwards by centrifugal force. This movement is controlled by the escapement mechanism and limited by a stop pin. The percussion rod is then supported by the creep resulting from deceleration.

On impact the percussion rod is driven in by the hammer. The thrust of the rod on the inclined projection of the spring catch disengages the catch from the centrifugal plate which is then free to move outwards taking with it the base end of the firing arm. This movement of the firing arm sets free the arm on the retaining bolt and thus releases the retaining catch and the detonator pellet.

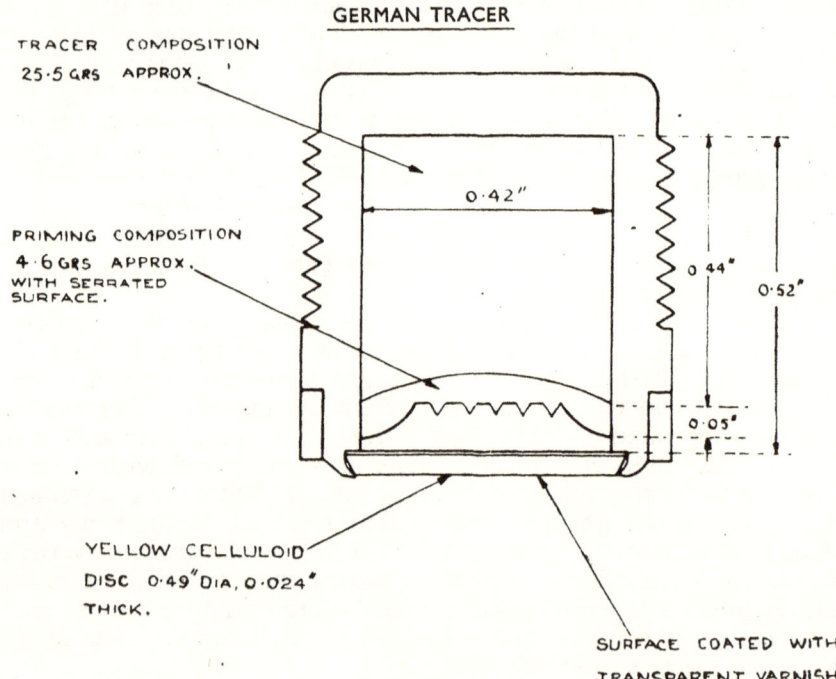

FIG. 11

# GERMAN 4·7 cm. PAK H.E. SHELL

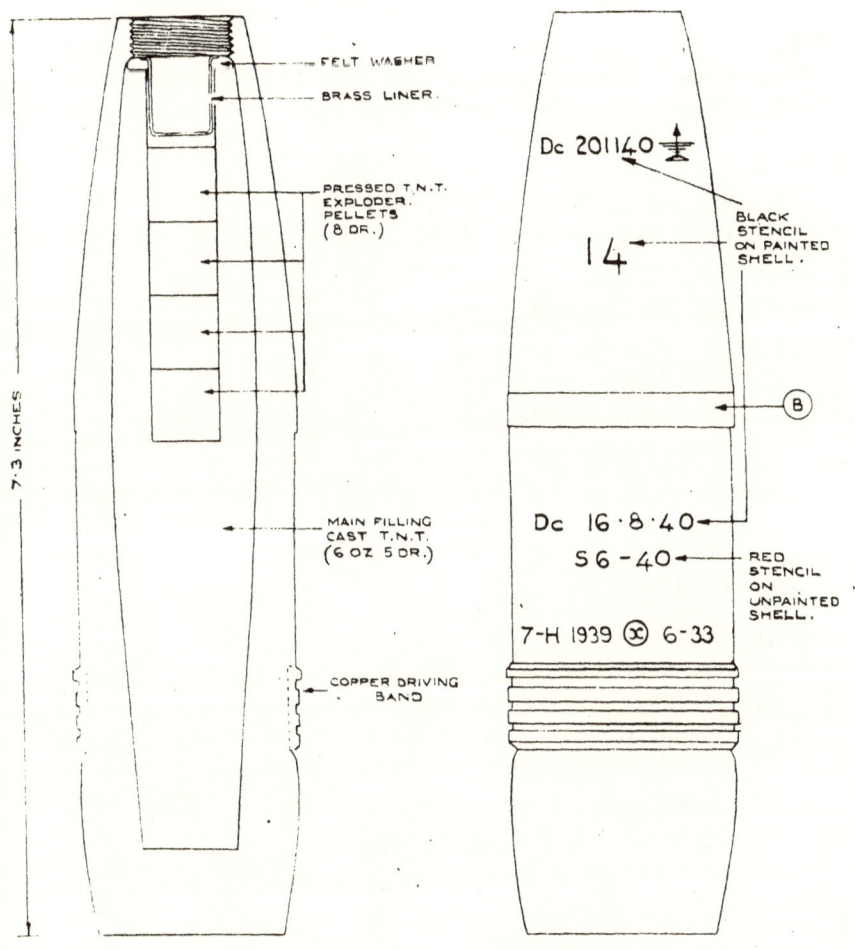

Fig. 12

# GERMAN TRACER FOR 5 CM. A.P. 40 SHOT
### (Fig. 11)

The tracer described is fitted to the A.P. 40 shot fired by the 5 cm. Pak 38 anti-tank gun. It is similar to that employed with the 4·7 cm. A.P. 40 shot.

## Body

The tracer body is of rustproofed steel and is threaded externally for attachment to the tracer cavity of the shot. There are no markings.

## Filling

The two part filling is made up as follows :—

Tracing Composition. About 25·5 grains of the following compositions :—

| | | |
|---|---|---|
| Magnesium ... ... | 32 | per cent. |
| Barium Oxalate ... | 21 | ,, |
| Strontium Nitrate ... | 14 | ,, |
| Potassium Nitrate ... | 19 | ,, |
| Organic Matter ... | 14 | ,, |

Priming Composition. About 4.6 grains of a mixture of magnesium powder, barium and potassium nitrates, and strontium peroxide. Accurate quantitative figures are not available. The whole is closed by a celluloid disc.

## Colour and Duration

The colour of the trace is yellow, and its duration at rest is 3·2 seconds. It is estimated that the duration in flight would be between 2 and 2·5 seconds.

# GERMAN 4.7 CM. Pak(t) AND Pak. K.36(t) CARTRIDGE, Q.F., H.E.
### (4.7 cm. Pak. Spgr.)
### (Fig. 12)

One type of this 4·7 cm. fixed Q.F. round is described in Pamphlet No. 4, pages 16 and 17. Details are now available of another type of this round, which is of Czech origin.

## Markings and Weight

The shell is known both wholly unpainted, or in the standard German olive green shade. These markings are indicated on Fig. 12. The total weight is 6 lb. approximately.

## Shell

The shell is of forged steel the nose being formed by the "bottling" process. The cavity is unmachined except for an internal thread provided at the nose to receive the fuze. The driving band is wholly of copper, and no base plate is fitted. The total weight of the filled and fuzed shell is 3 lb. 5 oz.

## Bursting Charge

The bursting charge consists of 0·39 lb. of cast T.N.T. having a central exploder cavity.

## Exploder

The exploder consists of 4 pellets of pressed T.N.T. of total weight 8 drs. fitting into the cavity of the filling.

## Fuze

The fuze is of the type described in Pamphlet No. 4, page 17.

## Cartridge Case

The cartridge case is similar in all respects to that described in connection with the 4.7 cm. A.P.C. shell in this pamphlet.

## Propellant

The propellant is similar in constitution to that described in connection with the 4·7 cm. A.P.C. shell in this pamphlet but differences in charge weight and ballistic size are as follows :—

Charge weight    242 grams.       — 8 oz. 9 drs. approx.

Size       ...    (0·6 × 6 × 320) mm. — (0·022 × 0·24 × 12·6) inches.

## Primer

The primer No. Vz. 33 was fully described in Pamphlet No. 9.

## GERMAN 4.7 CM. PAK(t) AND PAK K.36 (t) CARTRIDGE Q.F., A.P.C. OF CZECH ORIGIN

(Fig. 13)

The fixed Q.F. round is fired from the 4·7 cm. anti-tank guns of Czech origin. The length of the complete round is 20·5 inches and the weight, 6 lb. 14 oz. 6 dr. The penetrative cap, which is black, is tapered towards its nose where it is circular and rises slightly to a point at its centre. The body is not painted and only a narrow strip of the driving band is visible at the mouth of the case. The letters " T.T." are stencilled in red on the body to indicate a bursting charge of pressed T.N.T. The serial and filled lot markings are also stencilled in red, below the " T.T." marking. The brass cartridge case is approximately 16 inches in length and is very slightly necked. The designation " 4 cm. VZ 36 " is stamped in the base of the case which is also stencilled " 1/10 320 P.P. VZ. 36 " in black. These markings indicate the size, shape and nature, respectively, of the propellant.

Rounds of more recent manufacture have the shell, as well as the cap, painted black and red stencilling in accordance with the German

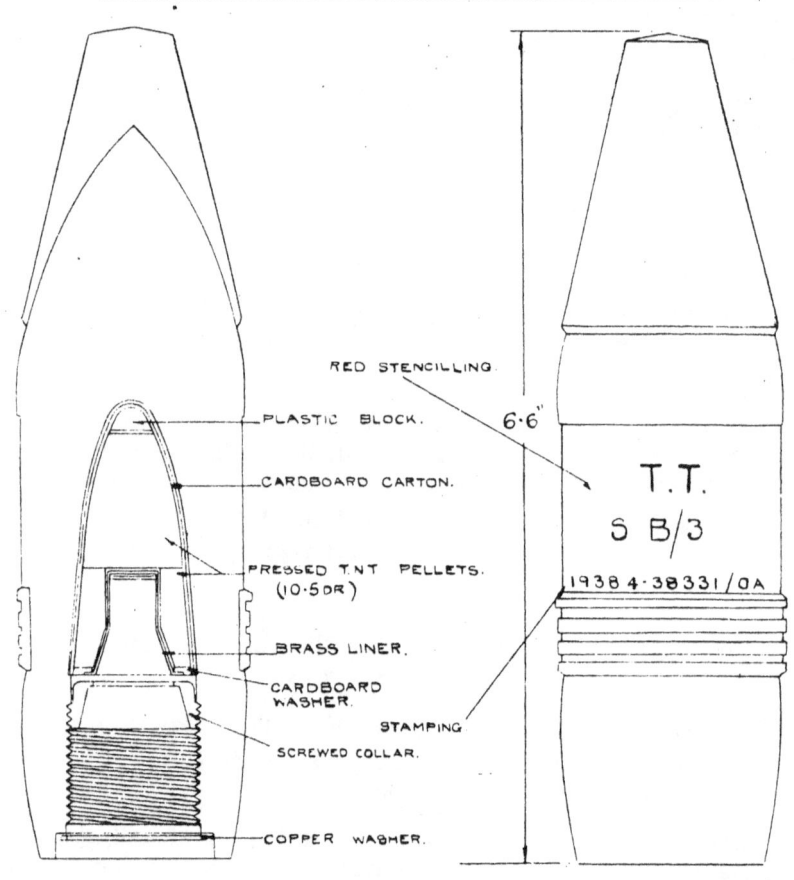

GERMAN 4.7 cm. (t) & PAK. 36 (t) A.P.C. SHELL OF CZECH ORIGIN

FIG. 13

system. With these rounds, the case is stencilled on the side in black to indicate the equipments, the weight of the propellant charge, the size, shape and designation of the propellant and places and dates of manufacture and filling. The following is a typical example of the stencilling : " 4·7 cm. Pak (t), 4·7 cm. Pak K.36 (t), 436g, 1/10/320 Str.P M36, sem. 1940/3, pla 26.5.5.1941 ". The usual red stencilling " P.T.+25° C.", indicating that the charge weight is based on a standard charge temperature of 25° C., is found above the flange. The German abbreviation for piercing projectiles, " Pzgr ", is stencilled in white on the base of the case.

The complete round consists of the following components :—

(a) Armour piercing shell fitted with a penetrative cap and filled T.N.T.
(b) Base fuze fitted with a gaine.
(c) Brass cartridge case, " 4 cm. VZ 36 " or " 4·7 cm. M36 ".
(d) Propellant charge of double base composition in the form of strips and a decoppering agent.
(e) Long percussion primer, Vz 33 dv or M.40 dv.

## Shell

The steel body is machined from rolled bar and has a hardened head. The V.D. Hardness figure for the head averages about 850. The cavity for the bursting charge is small, leaving a practically solid head, and is closed at the base by the fuze which has an adapter screwed in above it and a copper washer in front of the flange. The copper driving band has three cannelures and is bigger in diameter in front of the first cannelure. When assembled in the case, the mouth of the case is coned into this cannelure and only the high diameter part of the band is visible.

The steel cap, which is hardened throughout, is machined from bar and is attached by a soldering process. The average V.D. Hardness figure is about 625.

The weight of the empty shell is 3·3 lb.

The bursting charge has a weight of 10·5 drams and consists of two pressed pellets of T.N.T. contained in a cardboard carton. The upper pellet is solid. The lower pellet is annular and contains a brass liner which receives the gaine fitted to the fuze. The carton is closed at the base by a cardboard washer and has a block of the bakelite type at the nose which occupies the small end of the cavity. A brass collar is screwed into the fuze hole below the bursting charge. The weight of the shell, filled and fuzed, is 3 lb. 10 oz. 7 dr.

## Fuze

Details of the fuze and its gaine are not yet available.

## Cartridge Case

The brass case is 15·93 inches in length and has a very slight shoulder, formed by an increase in taper, near the mouth. The primer hole is screwthreaded to a diameter of 0·66 inches.

## Propellant Charge

The charge consists of flexible strips of a light brown colour with a glossy surface. A paper label, attached to the charge, reads " 445g. 1/10/320 P.P. VZ. 36 S.4/36 ". These details refer to the charge weight, size of the strip in millimetres, shape and designation of the propellant and the lot identification marking. The dimensions of the strip in inches are :—length 12·6, breadth 0·39, thickness 0·04. The weight of the charge examined was 15 oz. 3 dr. The propellant consists of 64·37 per cent. of nitrocellulose (nitrogen content 12·3 per cent.), 29·55 per cent. of nitroglycerine and 6·08 of centralite.

The decoppering agent is a metal strip weighing 25·8 grains with the dimensions $3·9 \times 0·24 \times 0·014$ inches. The metal consists of 61·3 per cent. of tin and 38·7 per cent. of lead.

The stencilling " Str.P." on the side of the cases of rounds of more recent manufacture, is the German abbreviation for strip propellant. The original Czech designation of the propellant, " VZ.36 " has apparently been changed to " M36 ".

## Primer

A description of the long percussion primer, Vz 33 dv is included in Pamphlet No. 9.

### GERMAN 7.5 CM. Kw. K. CARTRIDGE Q.F., H.E.
(Fig. 14)

Examples of this type of ammunition have already been described in Pamphlet No. 4, page 31, and Pamphlet No. 7, page 54. Additional details are appended.

## Bursting Charge

In addition to the Amatol filled shell described previously a sample of this shell has now been examined which is filled T.N.T.

The weight of the filling, which is cast, is 1·5 lb. approximately. The nature of filling is indicated by the stencilled figures 14.

## Propellant

The tubular distance piece described in connection with the propellant charge (Pamphlet No. 7) has been found to be of the double based nitrocellulose—diethylene glycol-dinitrate type, of the following composition.

| | | |
|---|---|---|
| Nitrocellulose ... | 61·29 | per cent. (Nitrogen 12·95 per cent.) |
| D.E.G.D.N. ... | 36·13 | ,, |
| Potassium Sulphate | 2·22 | ,, |
| Graphite (coating) | 0·35 | ,, |

# GERMAN 7·5 Cm. K.w.K. H.E. SHELL FILLED T.N.T.

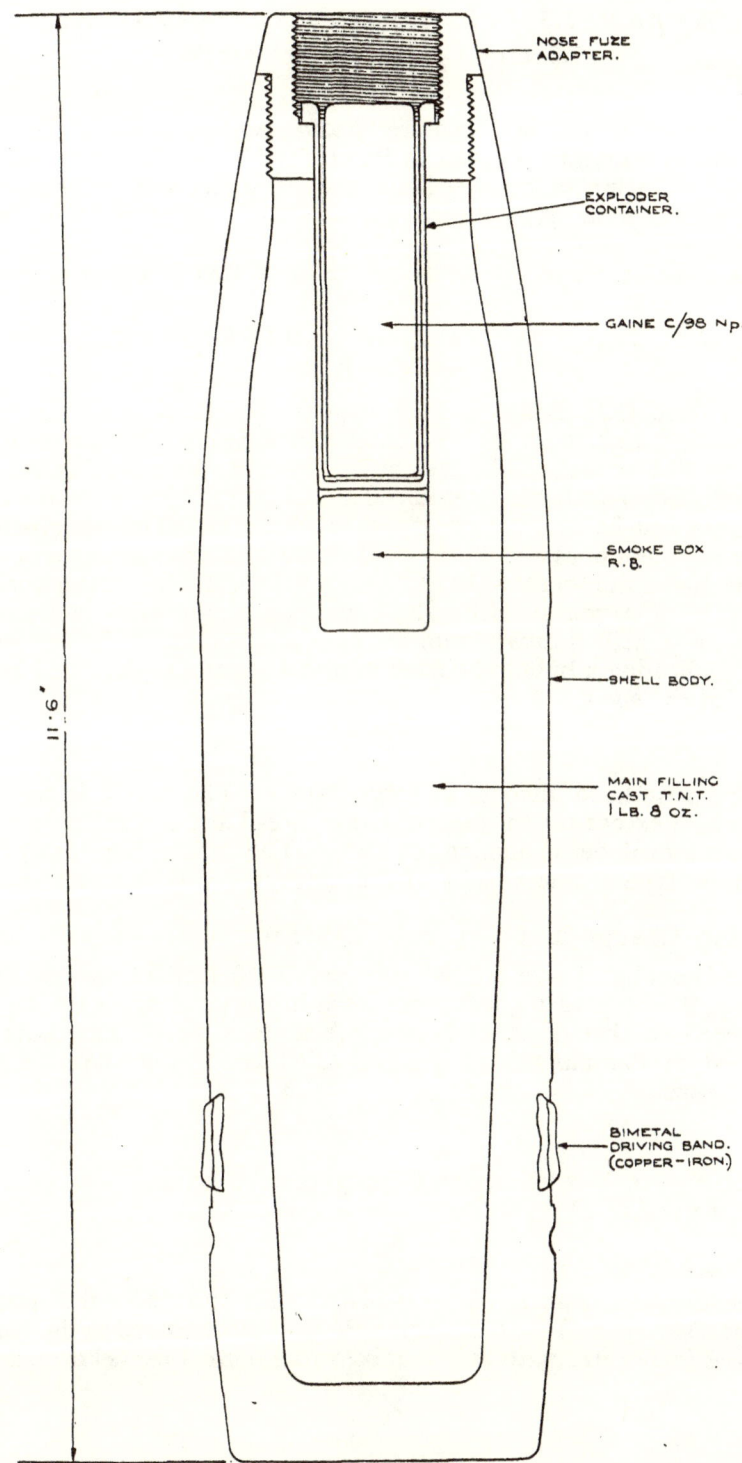

Fig. 14

## GERMAN 7.5 CM. L.G.40 Q.F. CARTRIDGE
### (7.5 cm. L.G.40 Hülsenkart)
### (Fig. 15)

A description of this separate loading Q.F. cartridge for the recoilless airborne gun is given in Pamphlet No. 8 where the case is described as being of brass. Since the pamphlet was published a critical examination has revealed the case to be of steel with brass plating.

The components of the cartridge are illustrated in Fig. 15.

## GERMAN 7.62 CM. Pak. 36 CARTRIDGE Q.F. H.E.
### (Figs. 16 and 17)

This fixed Q.F. round is used with the 7·62 cm. anti-tank gun (Pak. 36 (r)) and, with its 28·2 inch cartridge case, has an overall length of 38·8 inches. The weight of the complete round is 22 lb. The shell, fitted with the fuze kl.A.Z.23, is painted the normal deep olive green but has a 1·2 inch green or white tip. The designation of the equipment does not appear on the case which has stamped in the base the model number " 6340 St " and " Pak 44 Rh ". The marking " Sprgr ", indicating an H.E. Shell, and the weight class in Roman figures are stencilled in white on the base of the case. The numeral " 13 ", indicating a bursting charge of 40/60 Amatol is stencilled near the top of the shell.

### Shell

The streamlined shell is of forged steel. The cavity is wholly unmachined except for an internal screw thread at the nose to accommodate the exploder container and fuze. The driving band is of the bi-metallic type. The weight of the empty shell is 11·8 lb.

### Bursting Charge and Exploder System

The bursting charge consists of cast 40/60 Amatol and weighs 1·19 lb. The exploder system, as shown in the drawing, is similar in all respects to that of the 7·5 cm. Pak.40 H.E. shell filled Amatol described in Pamphlet No. 7. The filled and fuzed shell weighs 13 lb. 15 oz.

### Fuze

The kl. AZ.23 with optional delay of 0.15 second is a smaller model of the A.Z.23 type.

### Cartridge Case and Primer

The cartridge case is of brass-plated steel and bears the model number 6340 St and the abbreviation PAK 44 Rh stamped in the base. It is 28·2 inches in length and is affixed to a normal cannelure below

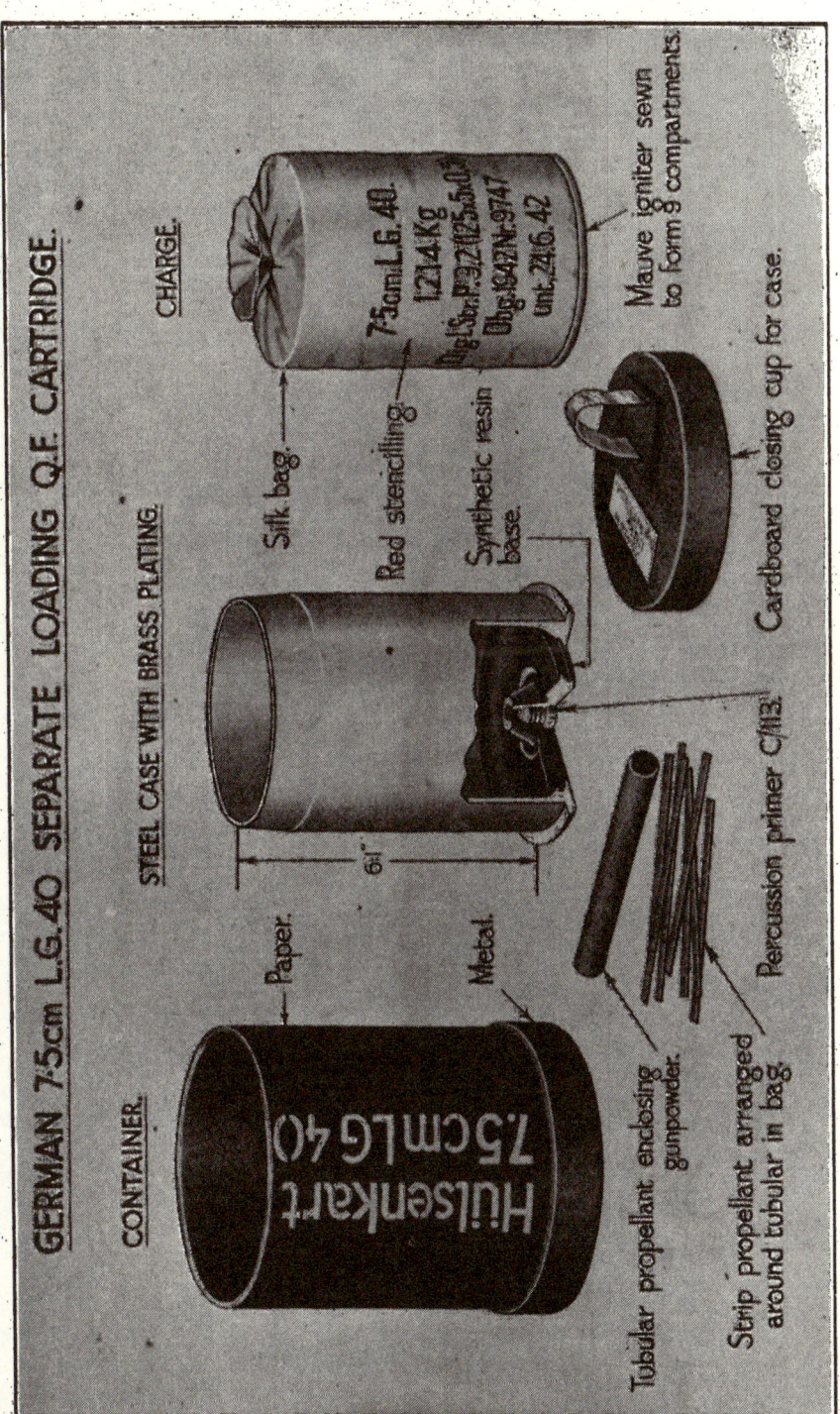

Fig. 15

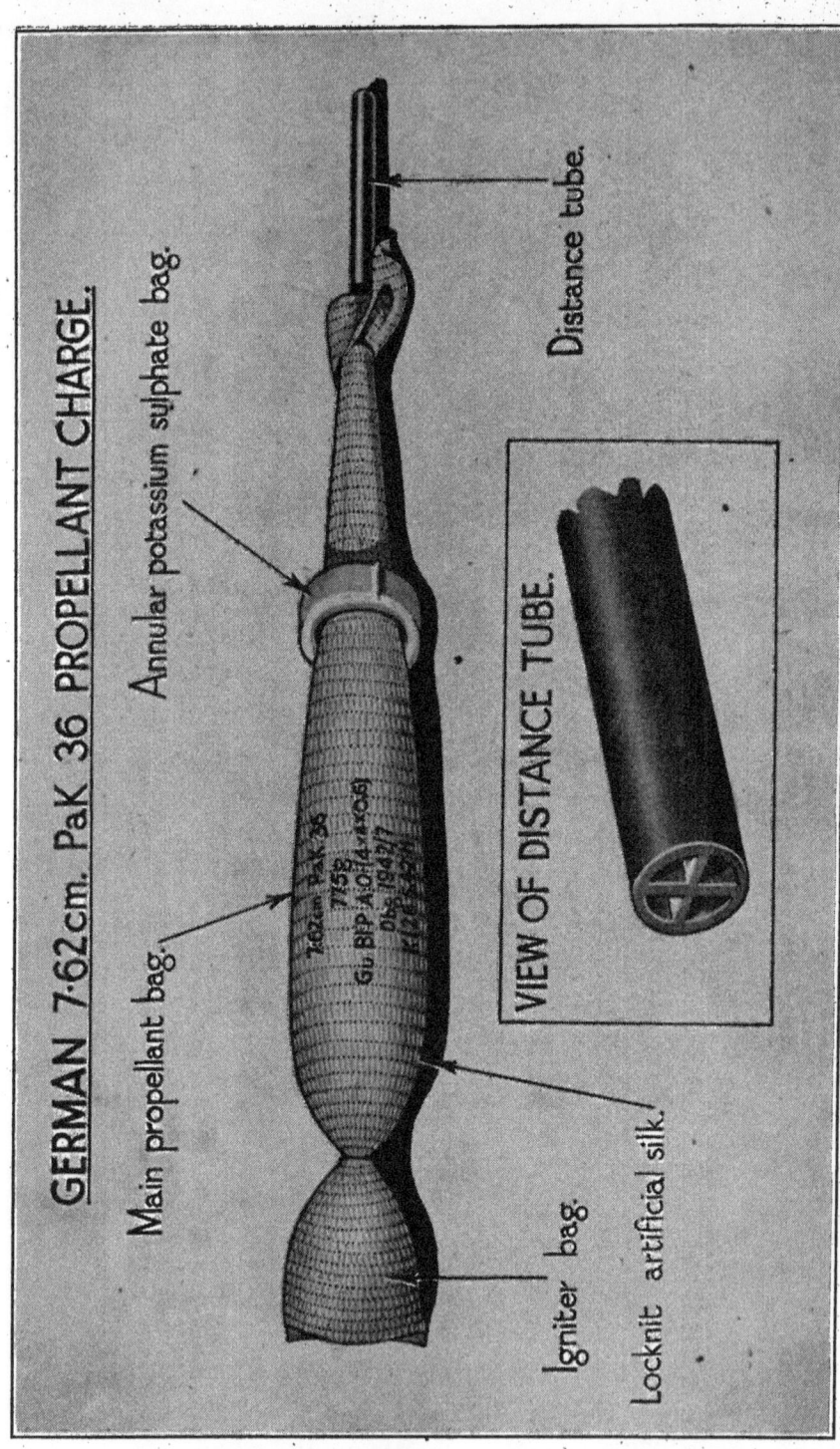

Fig. 16

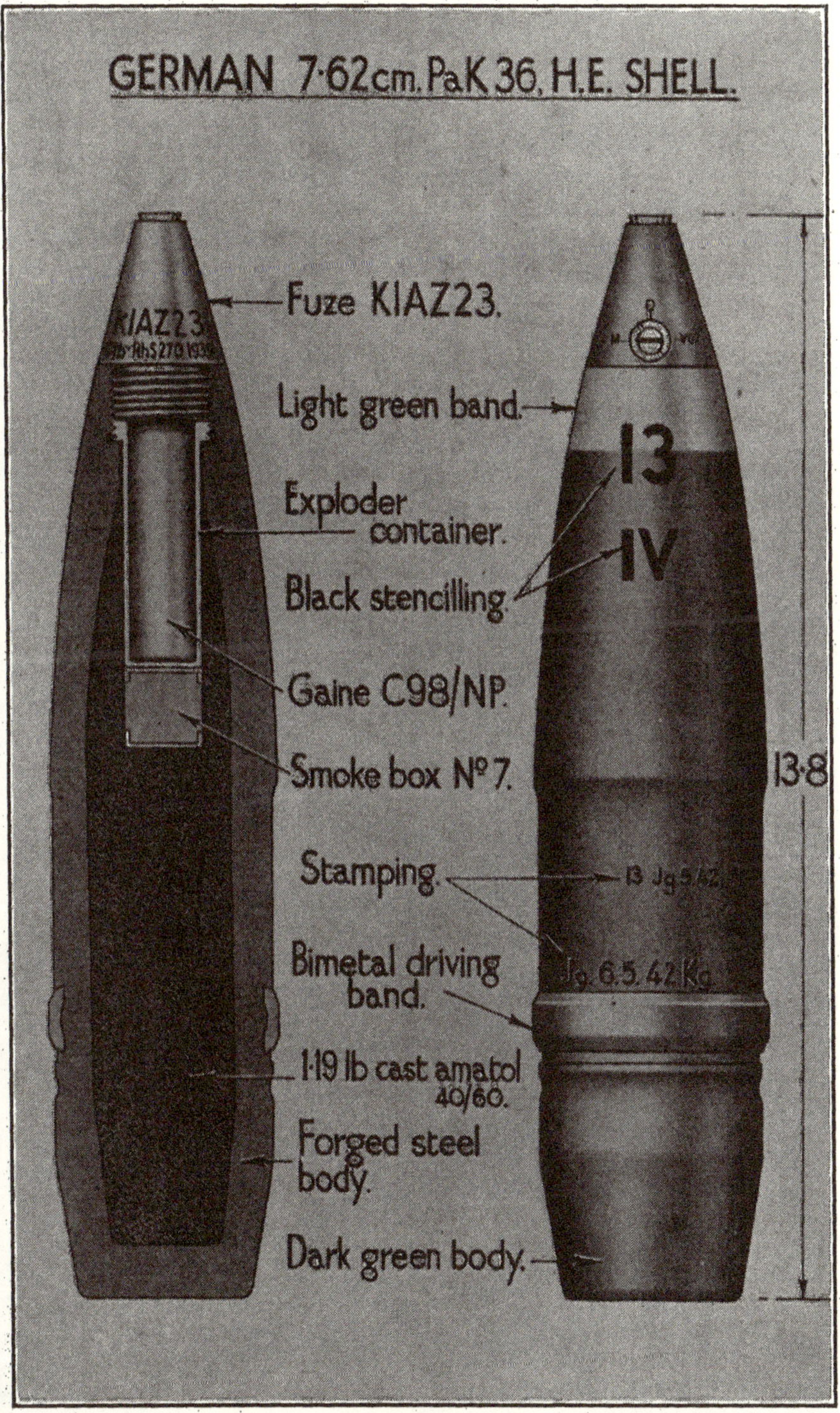
GERMAN 7·62cm. PaK 36. H.E. SHELL.
KIAZ23
Fuze KIAZ23.
Light green band.
13
IV
Exploder container.
Black stencilling.
Gaine C98/NP.
Smoke box Nº 7.
13·8
Stamping.
Jg. 6.5.42 Kg.
Bimetal driving band.
1·19 lb cast amatol 40/60.
Forged steel body.
Dark green body.

FIG. 17

the driving band of the shell. The primer C/12nA.St. is similar to that described in Pamphlet No. 4, page 10, but is of steel in lieu of brass.

**Propellant Charge**

The make up of the propellant charge is shown at Fig. 16 and will be seen to comprise four portions :—

1. The igniter pocket formed by sewing across the main bag which is of artificial silk. This contains about 1 oz. 13 drs. of nitro-cellulose powder in the form of cylindrical grains. The composition includes diethylelene glycoldinitrate (approximately 8 per cent.), centralite, diphenylamine, potassium sulphate and incorporated graphite.

2. The main propellant pocket, marked to indicate the weight, nature and size of the propellant, as shown in the illustration. This contains about 1 lb. 9 oz. of square flake propellant (Gu. Bl P.A.O. $(4 \times 4 \times 0,6)$, of the following composition :—

| | | |
|---|---|---|
| Nitrocellulose (Nitrogen 12·17 per cent.) | 38·6 per cent. | |
| Diethylene glycoldinitrate ... ... | 30·9 | ,, |
| Nitroguanidine ... ... ... ... | 30·2 | ,, |
| Graphite ... ... ... ... ... | 0·24 | ,, |

3. A central cross-webbed tube of propellant, of the form shown in the inset, which apparently acts as a distance piece for the charge. This has the following composition :—

| | | |
|---|---|---|
| Nitrocellulose (Nitrogen 12·91 per cent.) | 62·6 per cent. | |
| Diethylene glycoldinitrate ... ... | 34·8 | ,, |
| Potassium sulphate ... ... ... | 2·3 | ,, |
| Graphite ... ... ... ... ... | 0·3 | ,, |

The tube is 21·7 inches long and has a diameter of approximately 0·55 inch. The mean thickness of the cross-shaped web is 0·04 inch.

4. An annular bag marked "20g. $K_2SO_4$" which contains approximately 10 drams of potassium sulphate.

### GERMAN 10.5 CM. L:G.40 Q.F. CARTRIDGE
(Fig. 18)

The cartridge is used in the 10·5 cm. recoilless airborne gun (10·5 L.G.40, previously known as 1.G.2) and is a separate loading type with a plastic disc closing the base and a percussion primer fitted in the side of the case near the base.

The designation of the equipment is given on the label attached to the closing cup in the mouth of the case and is also imprinted in a circle on the plastic base. Only the earlier designation "1.G.2" or "L.G.2" has been found on cartridges examined to date.

The weight of the complete cartridge is 12 lb. 14 oz.

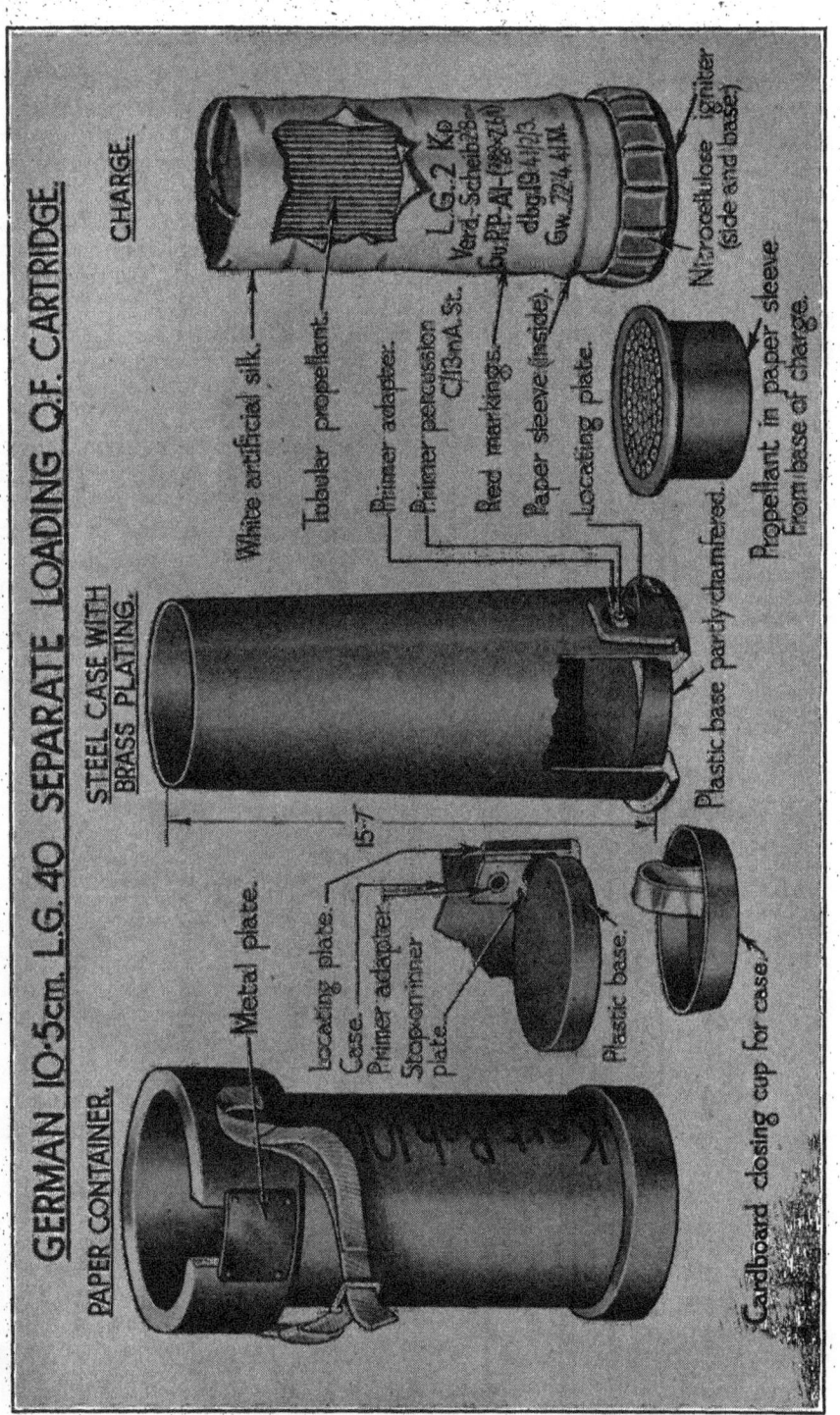

Fig. 18

### Case

The steel case, plated with brass, is 15·7 inches in length and tapers from 4·9 inches in front of the flange to 4·73 inches at the mouth. A large circular hole in the base has an internal flange on which a disc of brown plastic inside the base of the case is supported. A locating plate for the primer is fitted to the wall of the case near the base. The plate has a projection at the base which engages in a recess in the flange of the case and, near its forward end, is bored and screwthreaded to receive a screwed adapter for the primer. The inner end of the adapter engages a curved rectangular plate inside the case. This plate has a central hole for the adapter and, at its base, has a projection which engages a step formed in the plastic base piece.

The base piece is 4·1 inches in diameter and 1·1 inches in depth. The forward end of the cylindrical disc of plastic is chamfered along the part of its circumference which corresponds with the primer. The base is inserted with an adhesive and is located by a step at one end of the chamfer engaging the projection at the base end of the inner plate. Coinciding lines on the base of the plastic disc and under the flange of the case indicate the correct position for assembly.

The following variations in the construction of the case have been met with :—

(a) Locating plate integral with the body of the case.
(b) Absence of chamfer on base piece and of the projection on the inner plate.
(c) Absence of coinciding lines for assembly of chamfered base pieces.

### Primer

The percussion primer, C/13 nA St, used is described in Pamphlet No. 7.

### Propellant Charge

The charge is contained in a cylindrical bag of white artificial silk with an igniter at the base which also extends to form a collar around the lower end of the bag. The mouth of the bag may be choked or folded flat and secured by adhesive.

The 6-lb. 11-oz. 4¼-dr. propellant charge consists of a double base propellant (nitrocellulose and diethylene glycoldinitrate, including nitroguanidine) in the form of tubular cords, the diameter of the tubes being 0·102 inch externally and 0·039 inch internally. The charge is arranged in two bundles one above the other. The bundle at the base consists of 2·4 inch tubular cords contained in a paper cylinder which has an external flange at the top. The tubular cords in the upper bundle are 11 inches in length and are secured by ties. The weights of the lower and upper bundles are 14 oz. and 5 lb. 13 oz. 4¼ dr. respectively. The markings indicating the weight, nature and size of the propellant are shown in the drawing.

The igniter, in the form of a cap which fits over the base of the cartridge bag, contains 30 grams (approximately 1 oz. 1 dr.) of the nitrocellulose powder Nz Man N.P. (1,5 · 1,5). This is the powder normally used for igniters and consists of nitrocellulose with the addition of potassium nitrate and is in the form of cylindrical grains. The white material forming the igniter bag is stitched to form nine compartments for the powder at the base and twelve compartments in the band which fits around the lower end of the cartridge bag.

The charge is retained in the case by a cardboard cup inserted in the mouth of the case. The cup carries an adhesive label giving the same particulars as those marked on the bag.

## Packing

The cartridge is carried in a rolled paper container weighing 2 lb. 10½ oz. The approximate overall dimensions of the container are, length 16 inches, diameter 5·8 inches.

## GERMAN 10.5 CM. 1. F.H. 18 H.E. SHELL
### (F.H.Gr. 38 Stg.)
### (Fig. 19)

This shell is fired from the 10·5 cm. l.F.H. 18 gun-howitzer. The filled and fuzed shell weighs 33 lb. 1 oz.

## Shell

The streamlined shell is an annealed steel casting, machined externally overall. The cavity surface is unmachined except for provision of an internal thread at the nose to receive the exploder container and fuze.

The exploder container, which is 2·9 inches deep, consists of a drawn steel cup pressed into a threaded collar.

The single driving band is of the bimetallic type, no base plate is fitted. The empty shell and exploder container weigh 28 lb.

## Bursting Charge

The shell is direct filled with 3·8 lb. cast 40/60 Amatol. The central exploder cavity is 6·5 inches deep × 1 inch diameter and carries below the exploder container an 85 gram cardboard smoke box. This is unnumbered and contains 2 oz. 10 drs. smoke mixture of the following composition :—

|  |  |
|---|---|
| Red phosphorus ... ... | 85·4 per cent. |
| Paraffin wax ... ... | 11·6 ,, |
| Magnesium phosphates ... | 2·9 ,, |

## Fuze and Gaine

The A.Z.23 fuze used with this shell was described in Pamphlet No. 1, page 7.

The C/98 Np. gaine was described in Pamphlet No. 6, page 14.

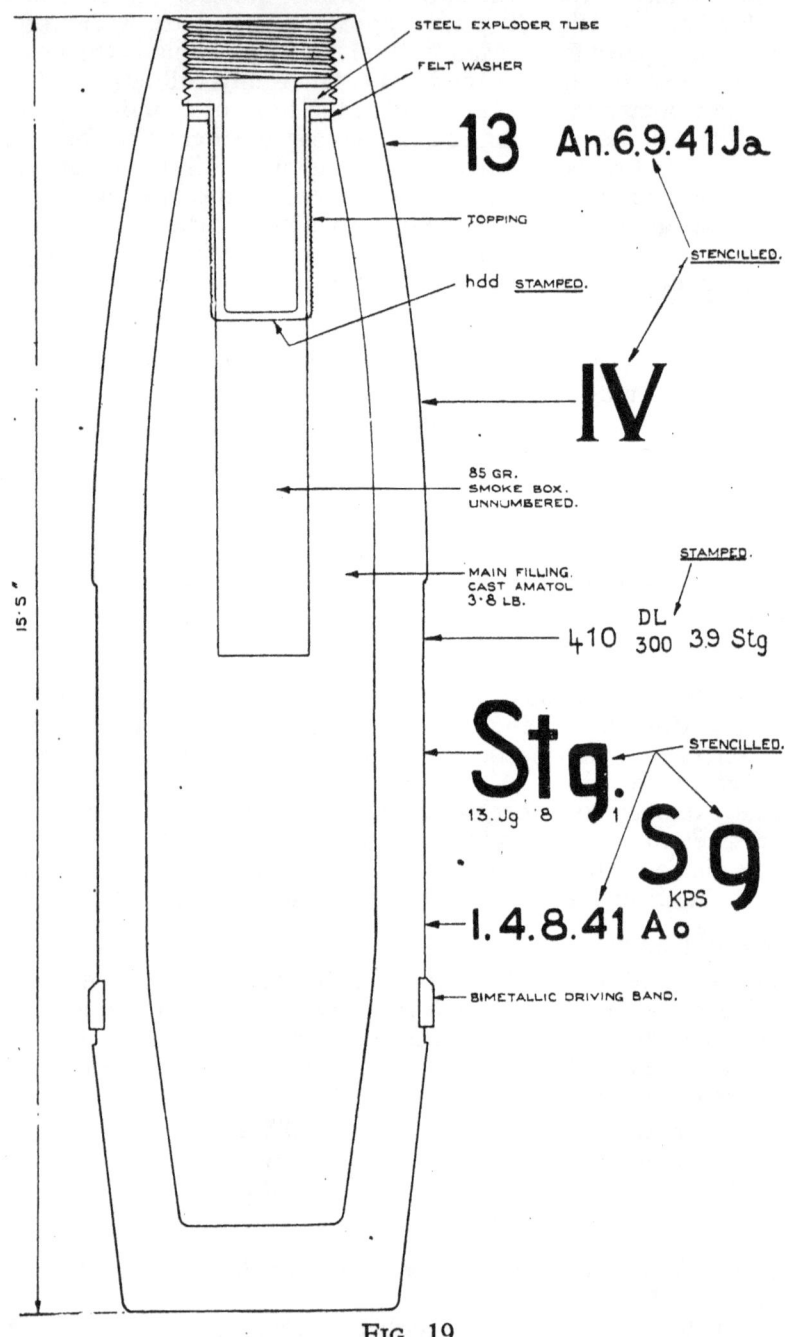

Fig. 19

# ITALIAN PROPELLANTS

## Nomenclature and Markings

There does not seem to be any rigid system of marking employed for Italian propellants. The markings which have been encountered may be divided into two groups, those invariably appearing and those appearing only intermittently. In the first group are included :—
  (i) The calibre and length (in calibres) of the weapon.
  (ii) Charge weight in grams or kilograms and propellant size in millimetres.
These may be the only markings found.
In the second group are included :—
  (i) The nature of propellant sometimes in full, sometimes abbreviated.
  (ii) A number, usually of seven figures, presumably a lot number and the date.
  (iii) Subsidiary markings such as contract numbers, manufacturers' marks, state and manufacturers' inspection marks. These appear variously on propellant bags, cartridge cases, or paper slips stuck to or merely inserted in the cartridges.
Comparatively few rounds which have been examined bear complete sets of markings.

## Composition

The composition of the propellants examined shows no outstanding features. There appear to be only three types in service :—
  (i) A straight ballistite (Ballistite or Bal.) containing about 50 per cent. nitrocellulose (Nitrogen 12·2 per cent.) and 50 per cent. nitroglycerine without stabilizer.
  (ii) A modified ballistite (Ballistite Attenuata or Bal. Att.) containing about 60 per cent. nitrocellulose (Nitrogen 12·6-12·8 per cent.), 26 per cent. nitroglycerine and 14 per cent. dinitrotoluene as a moderant coating.
  (iii) The so-called " Italian Cordite " which usually bears no markings other than the weapon and propellant sizes. This contains about 72 per cent. nitrocellulose (Nitrogen 12·8 per cent.), 24 per cent. nitroglycerine and 4 per cent. mineral jelly. In addition this type of propellant has been found to contain sodium carbonate or bicarbonate presumably as an additional stabilizer.
In some cases bags containing potassium chloride have been found included in the charges presumably as a flash reducing agent.

## Shape

Shapes encountered have been flat strip, cord, square flake, and tubular cord.

**Examples**

The following are examples of markings examined :—

(i) 65/17 Bal. 1 × 10 × 10 gr. 160—i.e. Ballistite for the 65 mm. 17 calibre gun in flake form size 1 mm. × 10 mm. × 10 mm. The charge weight being 160 grams.

(ii) BALLISTITE ATTENUATA IN PIASTRINE— 1 × 10 × 220 DAYS CKA VI grs 570, i.e. a charge of 570 grams of " Ballistite Attenuata " in strip form 1 mm. × 10 mm. × 220 mm. for the 75 mm. gun.

(iii) Typical of Italian practice for multi-part charges are the following markings found on the Q.F. separate cartridge for the 75 mm. 28 calibre gun/how.

1 Elem. Fondamen 77/28   i.e. lowest charge for the 77/28
1·5 × 15 × 15 kg. 0,224   gun/how.

2 Elem. Agguintivo da 77   i.e. Incremental charges for the
gr. 184   77/28 gun/how.
1·5 × 15 × 15

3 Elem. Agguintivo da 77
gr. 88
1·5 × 15 × 15

It will be noted (in this example) that the nature of the propellant is not given.

## ITALIAN H.E. HAND GRENADE OTO MODEL 35

(Fig. 20)

The grenade is of the percussion type with an " Always " action. The body is in two parts, a cylindrical cup-shaped base with rolled-on thread which mates with a similar thread rolled-on to the flange of the coned top. Two diametrically opposite slots are cut in this top immediately above the flange. When filled these two parts are secured by means of the locking spring shown in the diagram. The body is pressed aluminium painted red, and carries the safety cover drogue which is painted black. Protruding from the side of the safety cover is the end of the brass safety strip to which is attached a rubberized tab.

The bursting charge consists of about 2½ ozs. of low grade T.N.T. with a central detonator cavity. This is cast within a pressed metal cup. The cup is necked down at the top and is carried within the body of the grenade narrow end upwards. The detonator cap is pressed into a hole punched in the narrow end of the burster container, the lower end of this container being closed by means of a metal cover. Beneath the cap a flash sensitive detonator is inserted. Details of the composition of these detonators are not yet available.

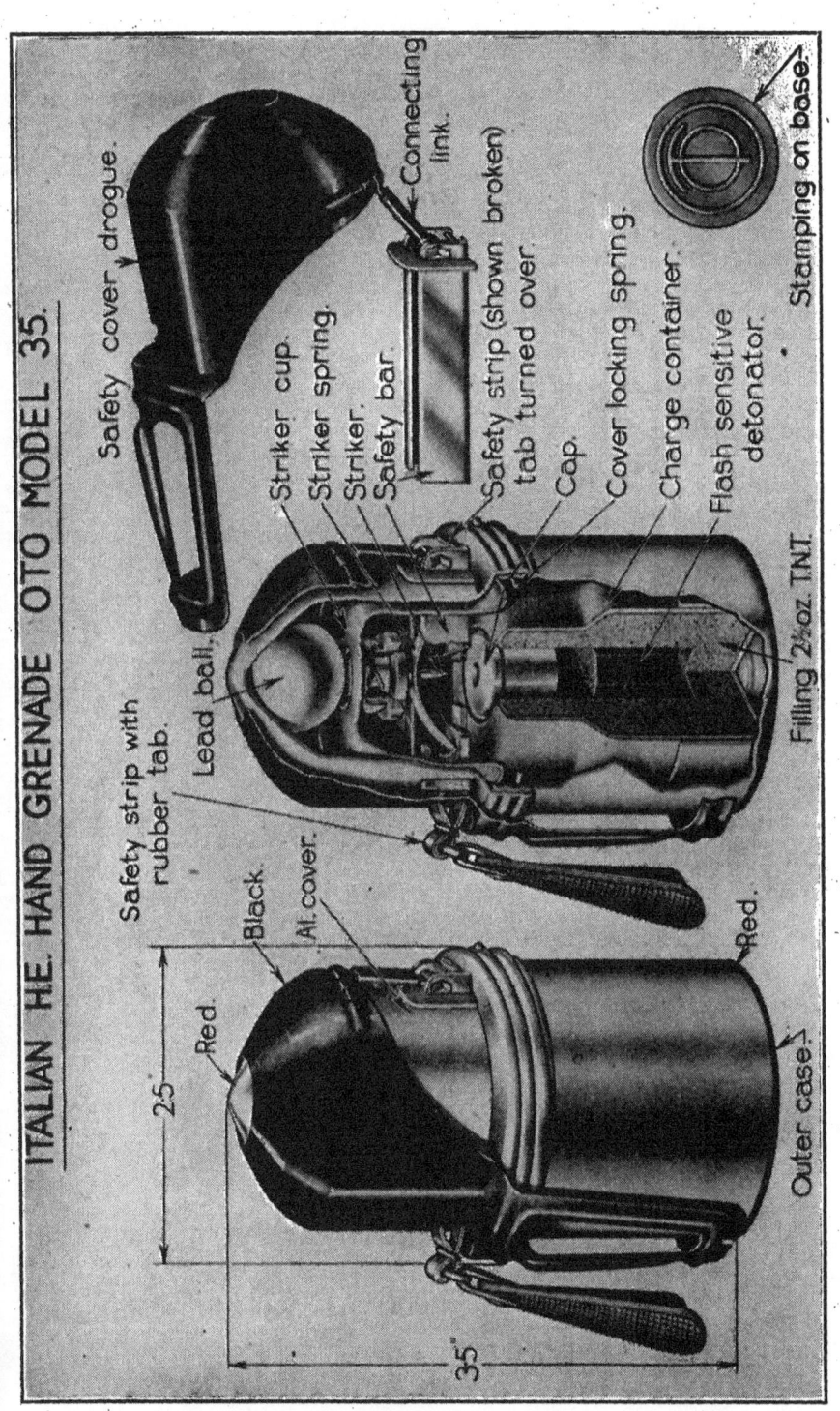

Fig. 20

**Fuze**

The " Always " fuze which resembles the British 247 fuze in its basic principles, is mounted within the coned cover of the grenade. It consists of a light aluminium cup fitting freely within the cover, a lead ball being carried between the cover and cup. Two diametrically opposite slots are cut in the walls of this cup to correspond with those cut in the outside cover. The striker is carried by an aluminium bush secured to the centre of the cup. Attached to the drogue by means of a light linkage is the safety-bar which comprises an open square section channel member.

In the " safe " position this safety-bar passes through the slots in the grenade cover, and striker cup, the open side being uppermost. The striker passes through a hole in the centre of the flat striker spring and the whole assembly is kept clear of the detonator by the closed side of the safety-bar. The drogue safety cover then folds over the top of the grenade and is secured in position by the brass safety strip which passes through two slots cut in its side, the lower longer portion passing beneath the safety-bar and being secured thereto by turning over on to a recess provided for that purpose.

**Action**

The safety strip is withdrawn and the grenade thrown in the normal manner. The safety cover opens and acting as a drogue withdraws the safety-bar. The striker spring cushions the striker assembly from the cap until impact. Upon impact the two portions of the grenade, i.e. the burster container and the ball loaded striker cup set into one another and thus actuate the detonating train and bursting charge.

## ITALIAN 25 GRAM. T.N.T. GAINE FROM 76 MM. 40 CAL. Q.F. H.E. SHELL

(Fig. 21)

This gaine bears the following markings stencilled on the brass case :—

AL 239 936
TEAPL 23 VEN
937

The gaine consists of a brass cylinder with a perforated cap secured by bayonet pattern clips.

This main brass container which is approximately 2·8 inches long overall and 0·8 inch in diameter contains a pressed pellet of milled T.N.T. weighing about 14 drs. and 2·6 inches in length. The container is coated internally with wax. A cavity about 1·1 inches in depth and 0·4 inch in diameter is pressed in the centre of the pellet.

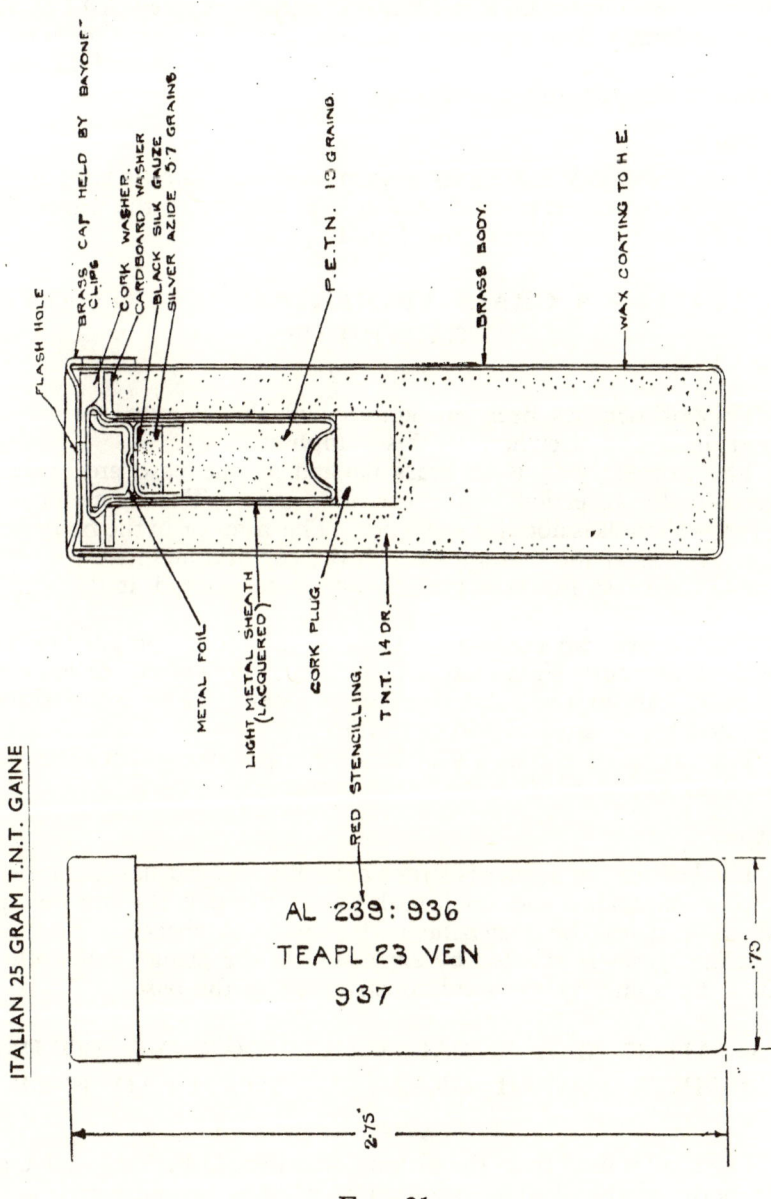

Fig. 21

**Detonator**

Inserted in the cavity mentioned above is a silver azide/P.E.T.N. detonator contained in a light metal sheath marked A.1239 about 0·8 inch long. The detonator is positioned between the brass cap of the outer case and the base of the cavity by a perforated cork washer at the top and a cork plug at the base.

**Action**

Flash from any igniferous system over the gaine passes through the perforations in cap and washer and detonates the silver azide/P.E.T.N. detonator and so the T.N.T. pellet.

## ITALIAN 9 GRAIN PRIMER PERCUSSION Q.F. CARTRIDGE
(Fig. 22)

This primer has been encountered in the 76 mm. 40 cal. Q.F. round described elsewhere in this pamphlet.

The primer body is of brass flanged at the base and threaded externally for insertion in the cartridge case. The diameter across the screw threads is about 0·44 inches. The interior of the body carries a small recess at the base into which the cap and anvil are secured by means of a brass fire-hole plug. The latter closed at the top by a paper disc.

The copper cap contains 0·3 grains initiatory composition composed of Mercury Fulminate 23 per cent., Potassium chlorate 41·5 per cent., Antimony Sulphide 35·5 per cent. The composition is protected by an unvarnished tin foil disc.

The magazine contains 9 grains of gunpowder and is closed by a varnished millboard disc over tissue paper.

**Action**

The base of the primer is struck centrally and crushes the initiatory composition against the anvil, flash passes through the fire hole plug and so initiates the magazine and propellant charge. Sealing of propellant gases is effected by expansion of the primer walls into the primer boss and by the solid construction of the base.

## ITALIAN 37 MM. 40 CAL. CARTRIDGE Q.F. COMMON POINTED WITH GRAZE ACTION HEAD FUZE
(Fig. 23)

The shell is fired from the 37 mm. 40 calibre Q.F. A.A./A. Tk. gun. The body of the shell is coloured pale blue, the nose red, and is stencilled as indicated in the diagram. The total length of the round is 13 inches, the brass cartridge case being 7·9 inches long. The total weight of the complete round is 2 lb. 13 oz.

ITALIAN 9 GRAIN, PRIMER, PERCUSSION, Q.F. CARTRIDGE

BASE STAMPINGS.

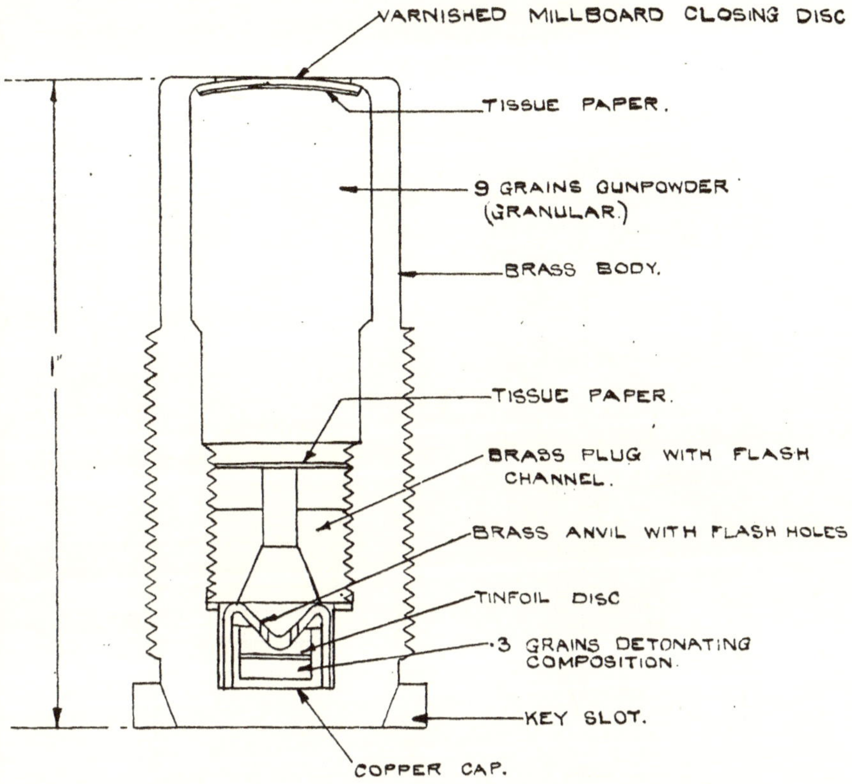

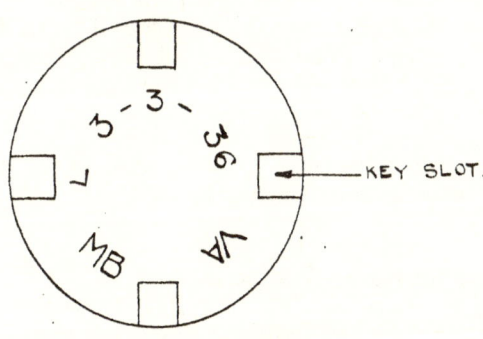

FIG. 22

## ITALIAN 37/40 COMMON POINTED SHELL WITH HEAD FUSE

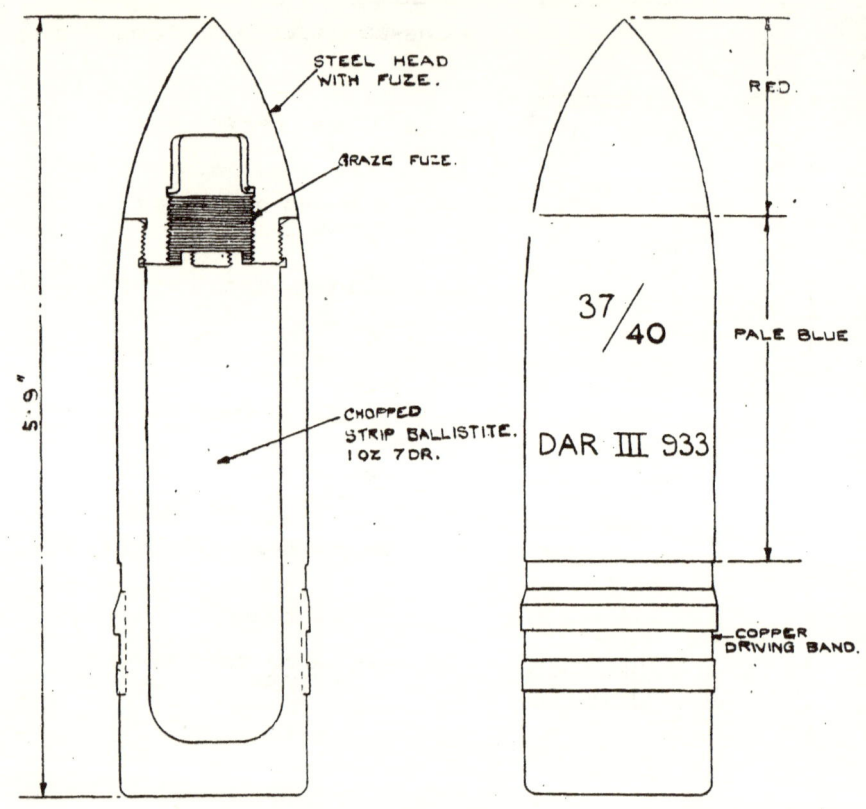

Fig. 23

### Shell

The shell is in two parts, a steel head drilled and tapped internally to take the fuze described below and threaded externally to fit into the comparatively thin walled body. The wide driving band is of copper. The weight of the filled and fuzed shell is approximately 1½ lb. The head is not hardened.

### Bursting Charge

The bursting charge comprises 1 oz. 7 drs. of chopped strip ballistite. There is no exploder system.

### Cartridge Case

The cartridge case is of brass 7·9 inches long, is fitted with primer Mod. 930 described in Pamphlet No. 9. The propellant charge is 3 oz. 5 drs. of the normal Italian ballistite described elsewhere in this pamphlet.

## Fuze
(Fig. 24)

The graze action fuze, which bears no Italian designation, screws internally into the steel head of the shell. It consists of a hollow brass body in the forward end of which an igniferous detonator is secured by a detonator plug of the usual design.

The hollow brass inertia pellet carrying a flat steel needle is positioned within the body by means of the brass split ferrule and the base plug. The inertia pellet and base plug are filled by perforated gunpowder pellets.

It has been found that a force of 45 lb. is required to force the inertia pellet through the split ferrule; this would be reduced in flight by the centrifugal forces acting upon the ferrule. It is not clear from available data whether this ferrule is designed to set back over the pellet during acceleration, thereby arming the fuze, or whether the pellet is designed to set forward through the ring upon impact.

Flash from the detonator upon impact is boosted through the perforated powder pellets, thereby firing the ballistite charge.

## Primer

The percussion primer, model 930, is described in Pamphlet No. 9.

# ITALIAN 37 MM. 40 CAL. CARTRIDGE Q.F. FUZELESS, SEMI ARMOUR PIERCING
(Fig. 25)

This round is used in the 37 mm. 40 Calibre Q.F. A.A./A. Tk. gun. The nose of the shell is coloured red, and the remainder of the body pale blue. It is stencilled as shown in the diagram. The total length of the round is 13 inches, the brass case being 7·9 inches long. The weight of the complete round is 2 lb. 14½ oz.

## Shell

The shell is steel and is apparently machined from bar stock, the cavity is closed by a left-hand threaded base plug and copper washer. The driving band is of copper. The shell is secured to the cartridge case by a continuous cannelure. The weight of the filled shell is 1½ lb. approximately. The V.D. Hardness figure for the head varies from approximately 700 at the point to 300 at the shoulder.

## Bursting Charge

The bursting charge consists of 13 drs. chopped flake ballistite. There is no exploder system and this charge is apparently intended to function by the heat generated upon impact.

## Cartridge Case and Propellant Charge

These are similar in all respects to those described for the common pointed round.

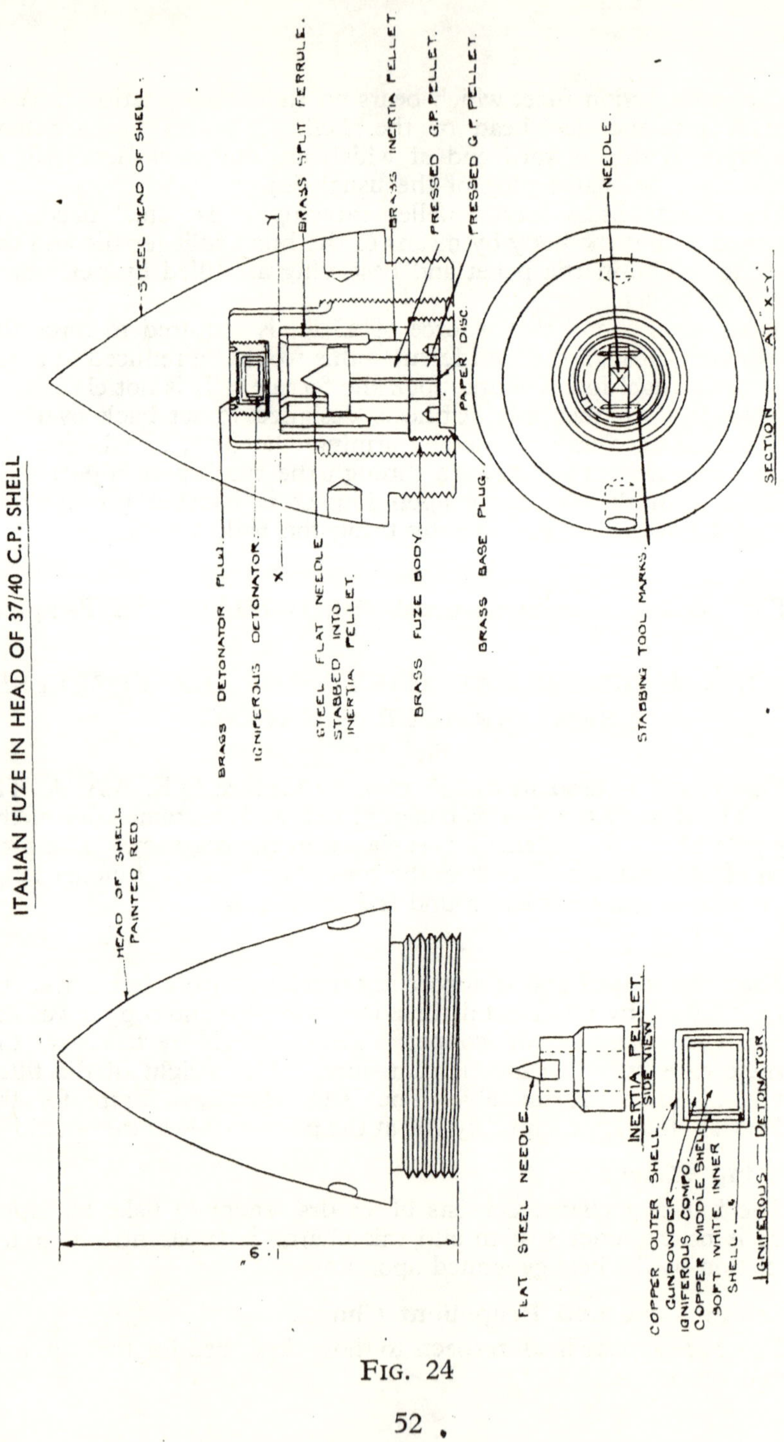

## ITALIAN FUZE IN HEAD OF 37/40 C.P. SHELL

**Labels (upper diagram):**
- STEEL HEAD OF SHELL.
- BRASS SPLIT FERRULE.
- BRASS INERTIA PELLET.
- PRESSED G.P. PELLET.
- PRESSED G.P. PELLET.
- BRASS DETONATOR PLUG.
- IGNIFEROUS DETONATOR. X
- STEEL FLAT NEEDLE STABBED INTO INERTIA PELLET.
- BRASS FUZE BODY.
- PAPER DISC.
- BRASS BASE PLUG.

**Section at X-Y:**
- NEEDLE.
- STABBING TOOL MARKS.

SECTION AT X-Y.

- HEAD OF SHELL PAINTED RED.
- 6"

**FLAT STEEL NEEDLE.**

**INERTIA PELLET SIDE VIEW.**

IGNIFEROUS — DETONATOR
- COPPER OUTER SHELL.
- GUNPOWDER.
- IGNIFEROUS COMPO.
- COPPER MIDDLE SHELL.
- SOFT WHITE INNER SHELL.

FIG. 24

ITALIAN 37/40 S.A.P. FUZELESS SHELL

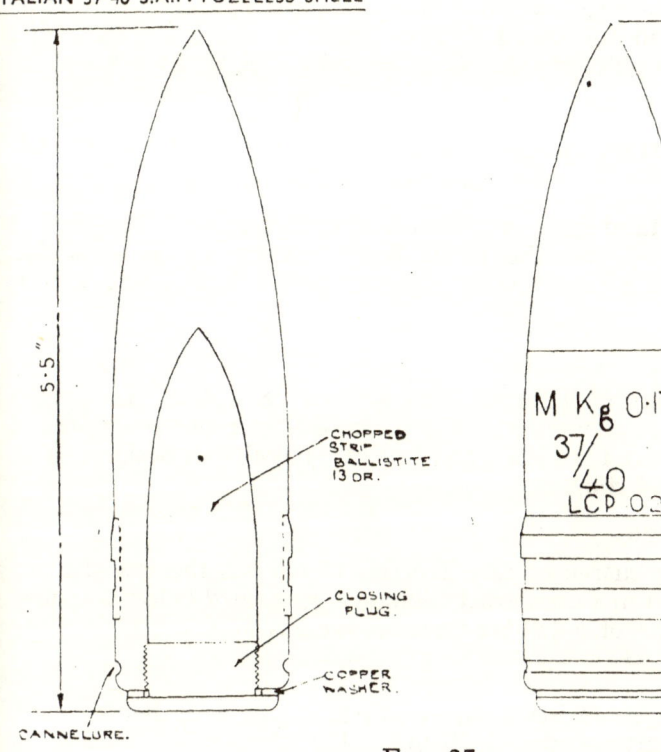

FIG. 25

**Primer**

The percussion primer, model 930, is described in Pamphlet No. 9.

### ITALIAN 65 MM. 17 CARTRIDGE Q.F. H.E.

One model of this fixed Q.F. round has been described in Pamphlet No. 7.

The following additional particulars are now available of an earlier model. This shell is coloured in exactly the same manner as has already been described but bears the markings :—

Kg 4·2<br>TRITOLO<br>ASSV 1932<br>65/17

It differs in other respects from the later version only in the method of filling, which in this case comprises a direct cast T.N.T. bursting charge of weight 9½ ozs. The exploder system comprises a ballistite gaine and exploder the latter contained in a cardboard tube within an aluminium sheath. The gaine is the 2·3 gram type described in Pamphlet No. 7. The exploder weighs 13 grams.

## ITALIAN 70 MM. 15 CAL. Q.F. SEPARATE H.E. SHELL
(Fig. 26)

This shell is fired from the 70 mm. 15 calibre How. The shell is coloured deep red over the whole body and carries no stencilled marking. It has a length when plugged of 9·3 inches and weighs 10 lb. 5 oz.

### Shell
The shell body is of forged steel without base plate, and weighs 6·9 lb. The nose adapter which is machined from bar stock weighs 2·2 lb. The cavity is of the usual Italian parallel wall design. The driving band is of copper.

### Bursting Charge
The bursting charge of cast T.N.T., which has the low density of 1·47, is carried in a cardboard container positioned below the nose adapter by means of a felt washer. It weighs 10·6 oz.

### Exploder
The exploder comprises a bakelized paper tube with metal closing caps and contains 14 drs. ballistite. The density of loading is approximately 0·96.

### Fuze and Gaine
The fuze is of the Inneschi type used over a 2·3 gram ballistite gaine both of which have been described in Pamphlet No. 7.

## ITALIAN 76 MM. 40 CAL. CARTRIDGE Q.F. H.E.
(Fig. 27)

This fixed Q.F. round is used with the Italian 76 mm. 40 calibre A.A. gun. The base of the shell and body forward of the driving band are coloured cream. A narrow red tip is painted at the nose of the shell. The body to the rear of the driving band is unpainted. The shell wall bears the stencilled markings :—

76/40

T.U.R.

A L 359/936

Similar markings also appear on the base of the shell. The shell examined was closed by a transit plug similar in appearance to those

## ITALIAN 70/15 H.E. SHELL

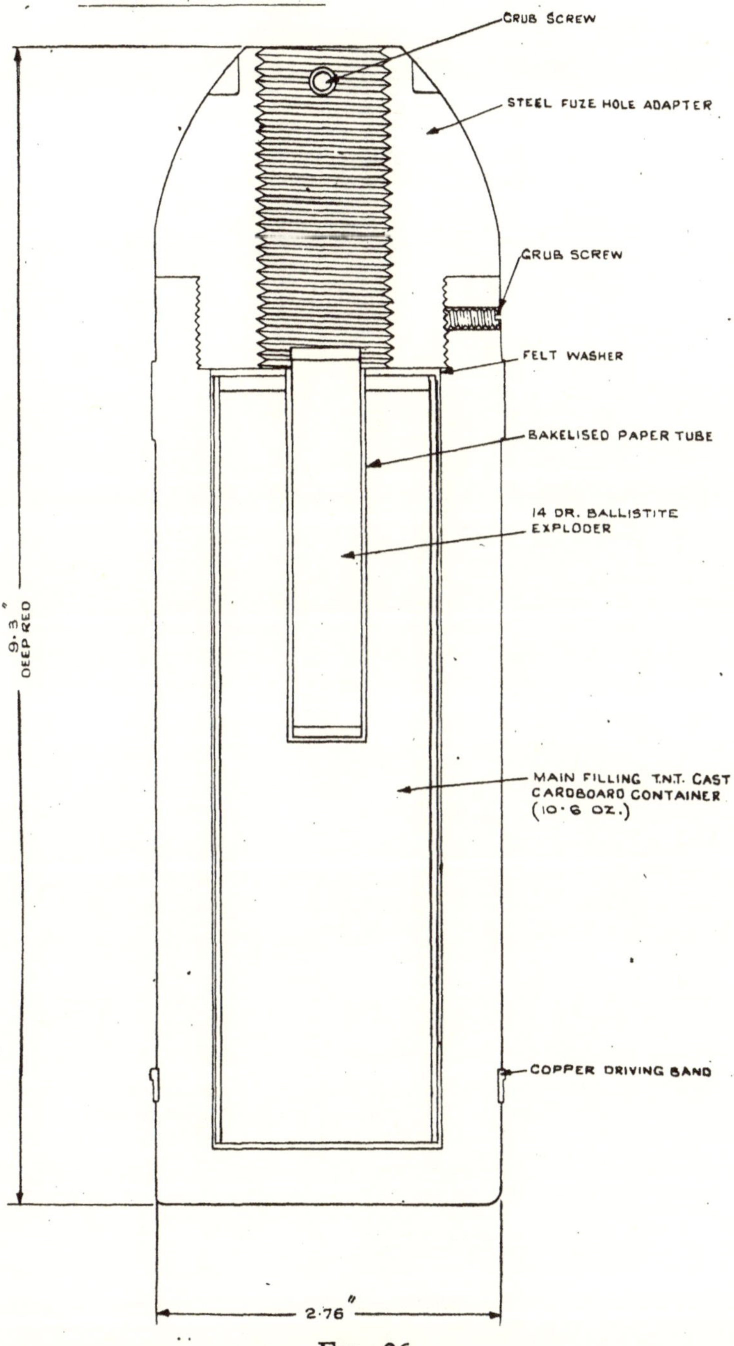

**Fig. 26**

Labels:
- GRUB SCREW
- STEEL FUZE HOLE ADAPTER
- GRUB SCREW
- FELT WASHER
- BAKELISED PAPER TUBE
- 14 DR. BALLISTITE EXPLODER
- MAIN FILLING T.N.T. CAST CARDBOARD CONTAINER (10.6 OZ.)
- COPPER DRIVING BAND
- 9.3" DEEP RED
- 2.76"

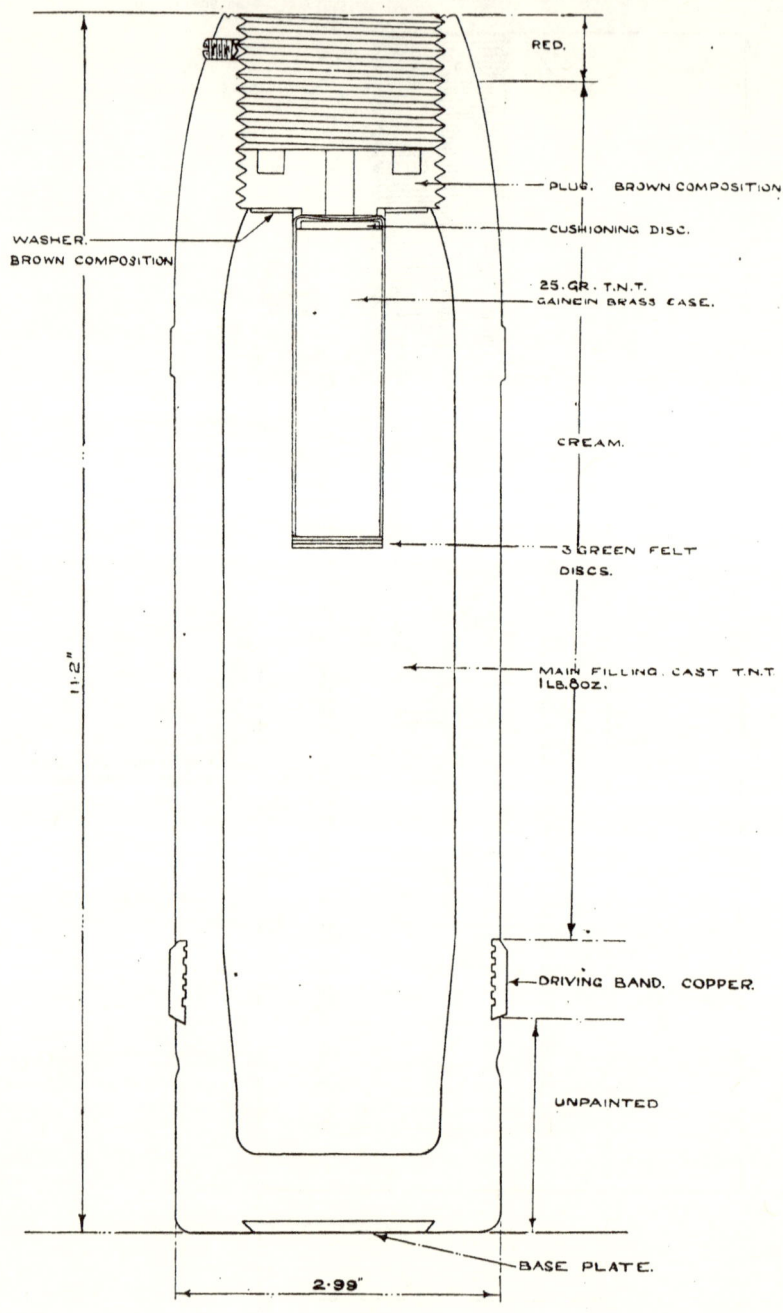

Fig. 27

employed in the British Service. The total length of the round is 26·6 inches and weight 20 lb. 9½ oz. A ring, one-half of which is blue and the other white, is painted under the flange of the 17·4-inch case.

## Shell

The shell is of forged steel throughout, the nose contrary to usual Italian practice having been formed by the "bottling" process. The upper portion of the cavity is threaded to take the fuze, and is otherwise machined only at the base end which is slightly coned. A pressed-in steel base plate is fitted. The single driving band is of pure copper. The plugged shell has a length of 11·2 inches. The weight of the empty shell is 11 lb. 3 oz. 8 dr. The weight, filled and plugged, is 13 lb. 6½ oz.

## Bursting Charge

The bursting charge consists of 1·48 lb. cast T.N.T. of melting point corresponding to British Grade I quality. A cavity about 1·9 inches deep is cast at the top of the filling. The remainder of the top is covered by a brown composition washer.

## Gaine

The 25 gram T.N.T. gaine is described elsewhere in this pamphlet. It is positioned within the cavity by means of 3 felt washers and the brown composition screw plug with central flash hole shown in the diagram.

## Fuze

No details of the fuze used with this shell are available.

## Cartridge Case

The cartridge case is of brass, bearing on the base the following marks stencilled in blue "CA-XXIII-4° V-3-937 VER-V-3-937" together with a green disc and blue triangle. It is secured to the shell by indentation into a continuous cannelure on the shell body.

## Propellant Charge

The propellant is wrapped in a fabric bag and consists of 1 lb. 10 oz. (approximately) of double base propellant in cord form. The cords have a mean diameter of 0·06 inch and length 14·4 inches. The propellant corresponds to the "Italian Cordite" described elsewhere in this pamphlet. Inserted in the bag with the propellant are three small packages each about 100 grains in weight containing Potassium Chloride presumably as an anti-flash agent. Tied to the base of the propellant bag is the igniter composed of about 1¼ oz. gunpowder of somewhat finer grain than British G.12. The base of this bag is coloured red. The whole is covered by a firmly secured pressed paper cup, probably to ensure good contact between the igniter and primer.

## Primer

This is fully described in this pamphlet, under the title 9 grain percussion primer.

## ITALIAN 77/28 LUNGA, CORTA AND 6 MIGL. H.E. SHELL, SHRAPNEL SHELL AND SEPARATE LOADING Q.F. CARTRIDGES

This ammunition is used in the Italian 77 mm. 28 calibre gun. Four types of shell have so far come to light, and following notes are divided thus:—

> Table of shell giving external detail.
>
> Methods of filling and relevant diagrams.
>
> Cartridge case and propellant.
>
> Table of propellant charges.

### 1. TABLE OF ITALIAN 77 MM. 28 CAL. SHELL

| Shell | MARKINGS | | Length | Remarks |
|---|---|---|---|---|
| | Stencil | Colour | | |
| H.E. LUNGA | Kg. 6,200 TRITOLO DAU M & L 1935 77 LUNGA | Head—Red Body—Light Blue Band above Dvg. band  Green | 11·2 ins. | H.E. method of filling as at (A) below and Fig. 28. |
| H.E. CORTA | Kg. 4,800 TRITOLO DAU (CML) 1935 77 CORTA | Head—Red Body—Light Blue Band above Dvg. band  Green | 9·6 ins. | H.E. method of filling as at (B) below and Fig. 29. |
| H.E. 6. MIGL. | Kg. 6,300 TRITOLO L.C.P. 1936 77/28 6 MIGL on reverse. | Head—Red Body—Light Blue Band above Dvg. band  Green | 11·2 ins. | H.E. method of filling as at (C) below and Fig. 30. MIGL indicates modified design—compare with above |
| SHRAPNEL | Kg. 5,600 77 FM D.A.Vr.IV. 1935 | Band at shoulder— Red Body—Light Blue Band above Dvg. band  Brown | 10·4 ins. | Shrapnel — method of filling as described below at (D) and Fig. 31. |

## 2. METHODS OF FILLING ITALIAN 77 MM. SHELL

### (A) 77 mm. H.E. Lunga (Fig. 28)

This shell has a sherardized metal surface and is coloured and marked as in the table above. The shell body is made of forged steel, and has no base plate. The cavity is slightly coned at the rear end. The exterior surface is machined to size, as is also the rear end of the cavity. The driving band is of pure copper. The nose fuze adapter is machined from bar stock and is threaded internally to carry the fuze and gaine and externally for attachment to the body.

The filled and fuzed shell weighs 13 lb. 13 oz. The weight of the empty shell is 11·7 lb.

### Bursting Charge

The bursting charge consists of 1 lb. 8 oz. cast T.N.T. carrying at the forward end a cavity for the exploder.

### Exploder

The 14 dram exploder consists of flake ballistite carried in a cardboard tube fitted in an aluminium sheath.

### Fuze and Gaine

The fuze is the Inneschi graze fuze used over a 2·3 gram ballistite gaine both of which have been described in Pamphlet No. 7.

### (B) 77 mm. H.E. Corta (Fig. 29)

This shell has a sherardized metal surface and is coloured and marked as in the table above. The shell is in three parts. The body is apparently of hot rolled tube and the nose has been forged to shape by the bottling process, it is threaded at the rear end to receive the base plug and the forward end to receive the nose adapter. The base plug is a machined drop forging and has a central cavity which carries the base of the filling. It is sealed by a copper washer. The nose adapter is machined from bar stock. Both external and cavity surfaces are fully machined. The driving band is pure copper. The filled and fuzed shell weighs 10 lb. 9 oz. approximately. The weight of the empty shell is 8·44 lb.

### Bursting Charge

The bursting charge is of cast T.N.T. and weighs 1 lb. 8 oz. approximately. A cavity about 3 inches deep and ¾ inch in diameter is cast at the top of the filling, and carries an aluminium sheath. The surface of the filling is covered by a thick felt washer.

### Exploder

The 14 dram exploder consists of chopped ballistite carried in a brass container and is positioned in the exploder cavity by means of a series of felt discs.

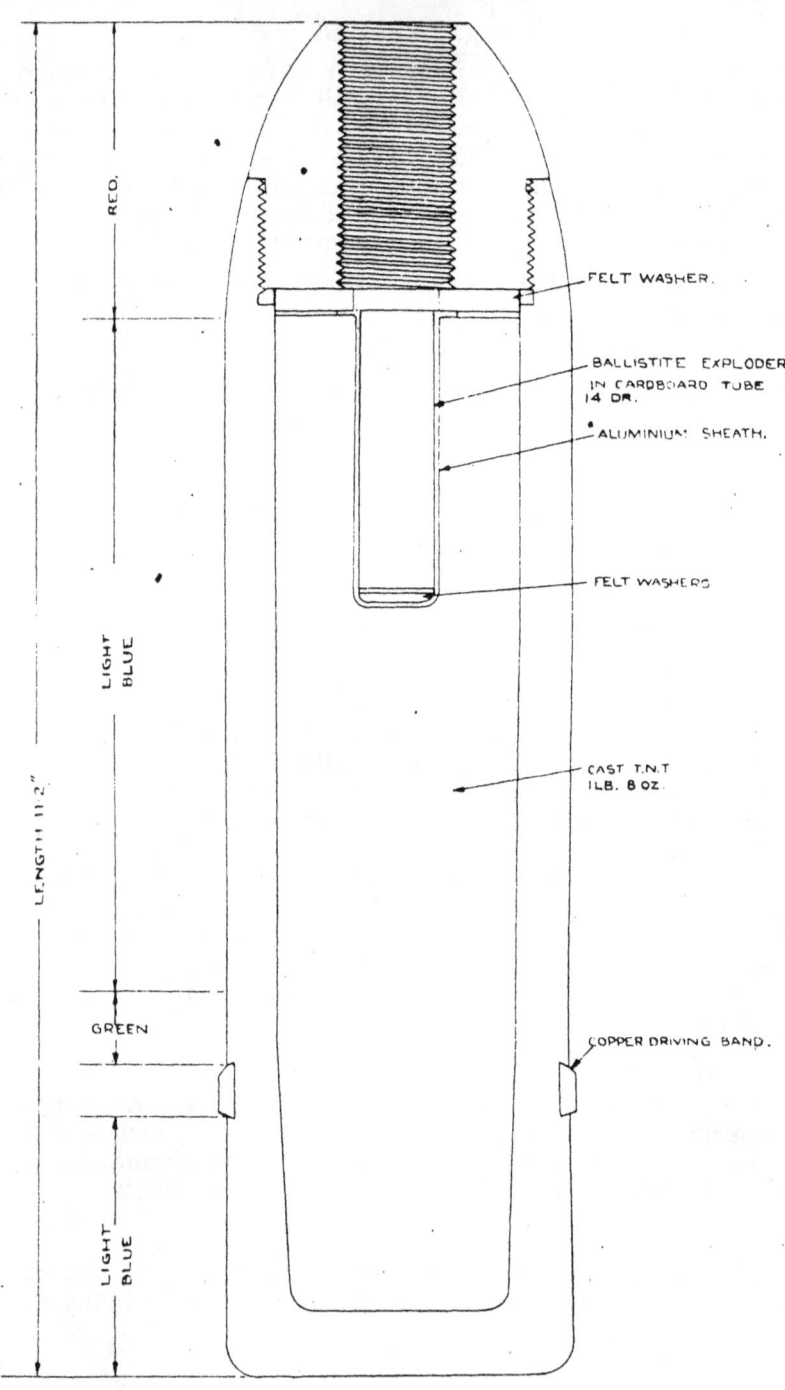

Fig. 28

# ITALIAN 77/28 H.E. SHELL "CORTA", METHOD OF FILLING.

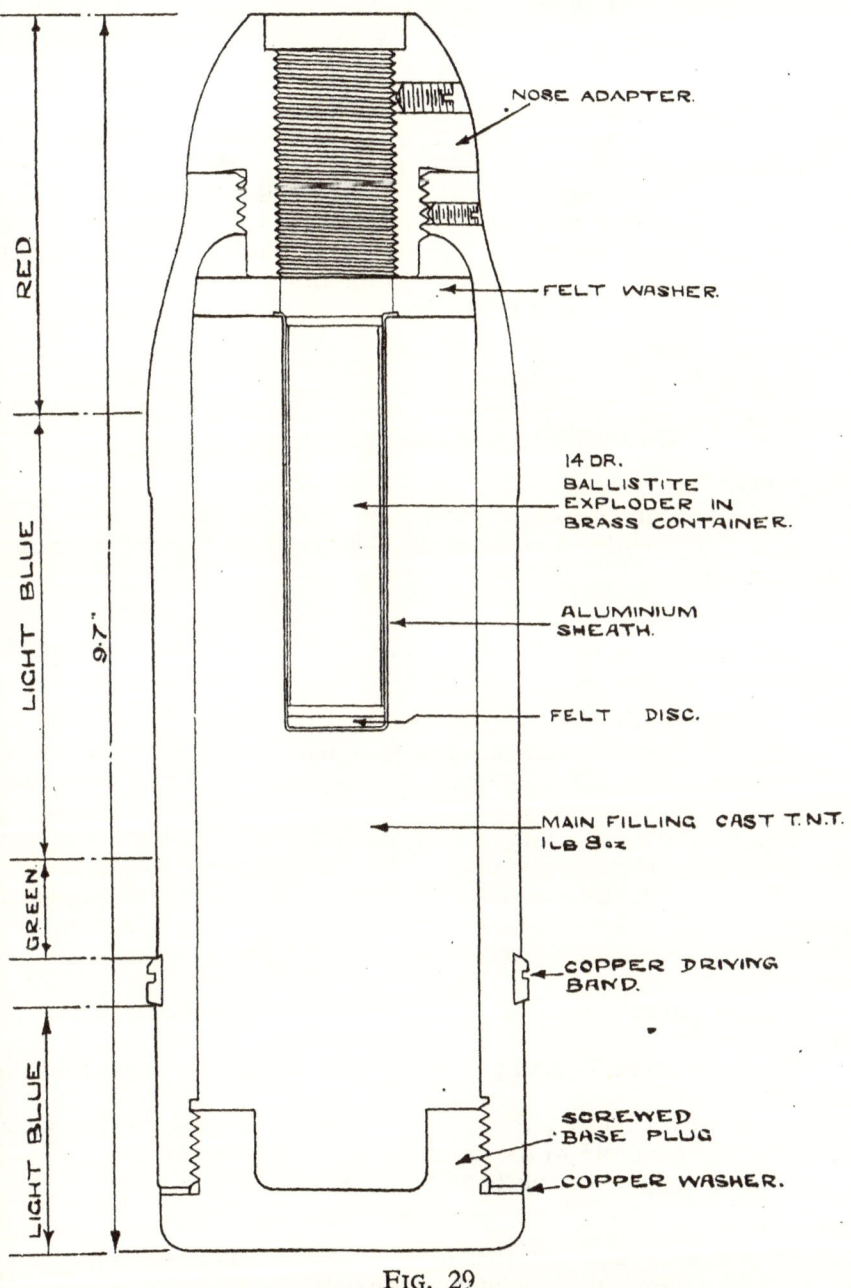

Fig. 29

### Fuze and Gaine

The fuze is of the Inneschi graze type used over a 2·3 gram ballistite gaine both of which have been described in Pamphlet No. 7.

### (C) 77/28 H.E. 6 Migl. (Fig. 30)

This shell has a sherardized metal surface and is coloured and marked as in the table above. The shell body in this case is forged of steel having a parallel walled cavity with a slightly coned base. The external surface and the radius of the base of the cavity have been machined. The shoulder is threaded internally to receive the nose adapter. A pressed in steel base plate is fitted. The nose adapter is machined from bar stock. The driving band is of pure copper. The filled and fuzed shell weighs 13 lb. 12 oz. approximately. The weight of the empty shell is 11·7 lb.

### Bursting Charge

The bursting charge is of cast T.N.T. and weighs 1 lb. 6 oz. and carries at the forward end a cavity for the exploder. The surface is covered by a felt washer.

### Exploder

The 14 dram. exploder consists of flake ballistite carried in a cardboard tube within an aluminium sheath, and positioned by means of felt washers.

### Fuze and Gaine

The fuze is of the Inneschi graze type used over a 2·3 gram. ballistite gaine both of which have been described in Pamphlet No. 7.

### (D) 77 mm. 28 Calibre Shrapnel Shell (Fig. 31)

The shell body is painted as has been described in the table above. It comprises the following components, nose adapter, flash tube, diaphragm and shrapnel bullets.

The shell body is of forged steel and is machined externally. The burster cavity and bearing surfaces of the diaphragm are also machined.

The nose adapter, which also forms a cover for the shrapnel bullets and provides a slot for the flash tube, is machined from bar stock. It is perforated centrally and has a depression for the initiating cap.

The flash tube is of brass and fits between slots in the nose adapter and diaphragm.

The diaphragm is stamped from steel plate and is perforated centrally.

The shrapnel bullets, about 180 in number, are of lead antimony alloy and are about 45 to the pound (0·48 inch diameter).

A single copper driving band is fitted.

### Method of Filling

An initiating cap. consisting of a 13-grain perforated powder pellet in an aluminium container, closed by paper discs, is inserted

ITALIAN, 77mm/28 cal. H.E. SHELL (6 MIGL)

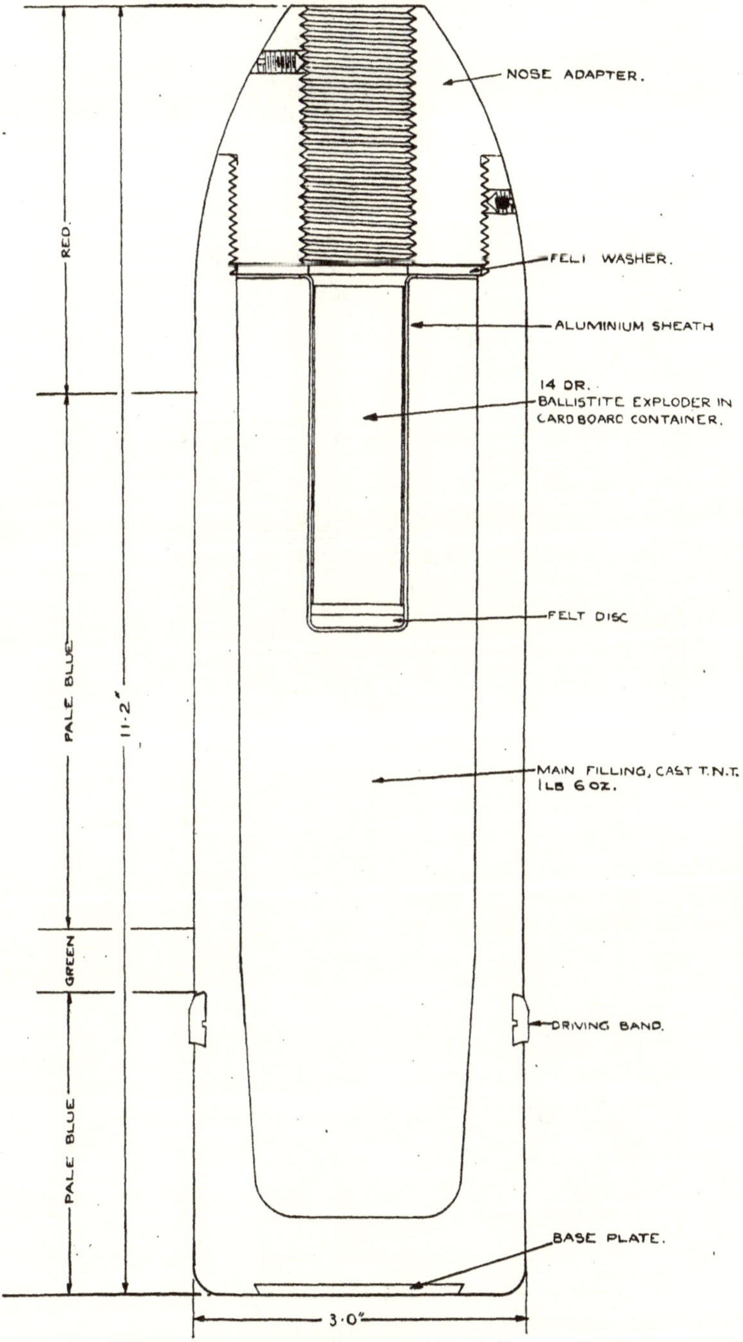

- NOSE ADAPTER.
- FELT WASHER.
- ALUMINIUM SHEATH
- 14 DR. BALLISTITE EXPLODER IN CARDBOARD CONTAINER.
- FELT DISC
- MAIN FILLING, CAST T.N.T. 1 LB 6 OZ.
- DRIVING BAND.
- BASE PLATE.

RED.
PALE BLUE
GREEN
PALE BLUE

11·2″
3·0″

FIG. 30

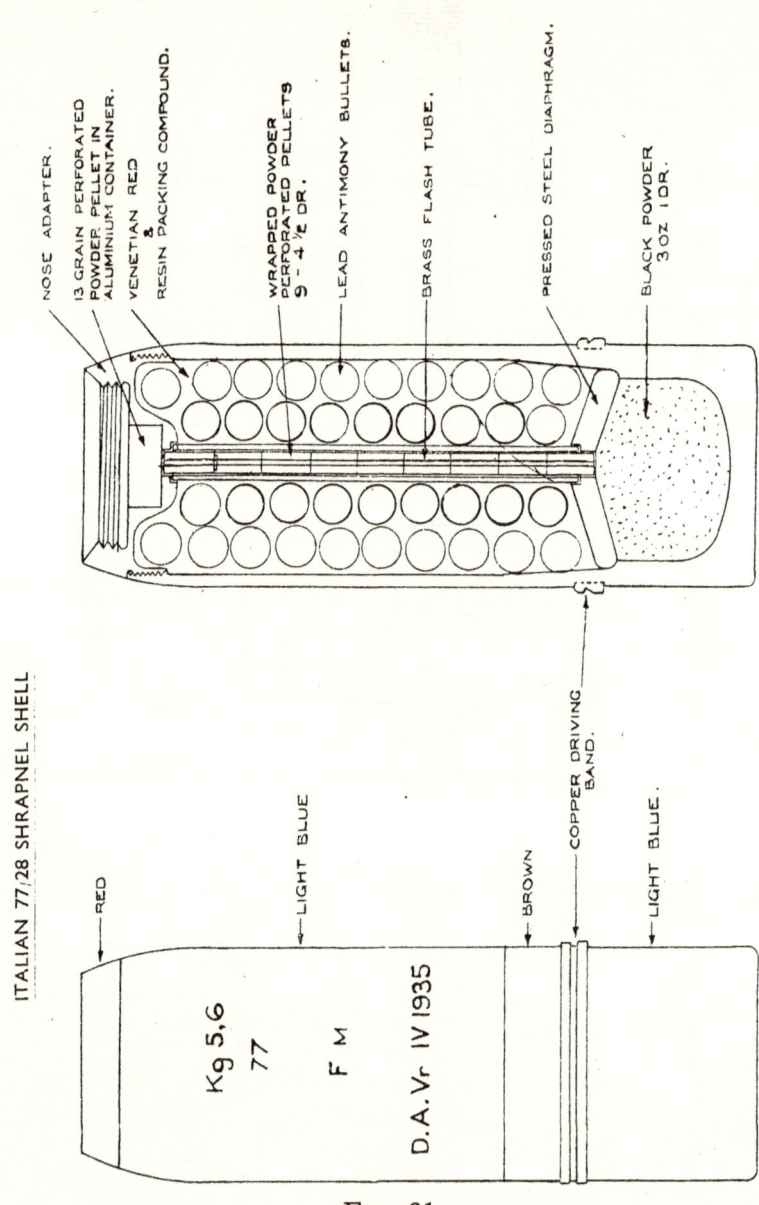

Fig. 31

in the nose adapter. The flash tube contains 9, 4½-drs. wrapped perforated powder pellets. (This device was employed in early British shrapnel shell to improve the dispersion of the bullets, but became obsolete early in the 1914-1918 war.)

The burster cavity contains 3 oz. 1 dr. black powder. The shrapnel bullets are packed around the flash tube by means of a composition of Venetian red and resin. The whole is covered by a resin topping.

**Fuze**

The time fuze graduated to 59 is fitted. This is fully described in Pamphlet No. 9.

### 3. Cartridge Case 77 mm. 28 Calibre

The cartridge cases so far encountered have been of brass and show the normal hardness gradients in the wall. The length is 9·2 inches, weight 2·1 lb. and capacity approximately 62 cubic inches.

Base stampings encountered sometimes include 77/28. It must be noted that 77/28 may not appear and only the old designation " 8 cm." may be present.

**Primer**

The primer MOD. 908 has been fully described in Pamphlet No. 9.

### 4. Propellant Charges

The charges are normally in three parts, except that the shell marked " CORTA " was associated with a two-part charge only. The propellant is contained in cloth bags marked as indicated in the table below.

Table 2. Propellant Charges 77/28 Gun

| Shell | Nature of Propellant | Increments of Charge | | | Remarks |
|---|---|---|---|---|---|
| | | Elem. " Fondam " (1) | Elem. Aggiuntivo (2) | Elem. Aggiuntivo (3) | |
| LUNGA | Ballistite | 224 gms. 1·5 × 15 × 15 | 184 gms. 1·5 × 15 × 15 | 88 gms. 1·5 × 15 × 15 | Square flake propellant. |
| CORTA | Ballistite | 224 gms. 1·5 × 15 × 15 | 184 gms. 1·5 × 15 × 15 | — | Two charges only. Propellant flakes as above. |
| 6 MIGL. | — | — | — | — | No details available. |
| SHRAPNEL | Ballistite | 224 gms. 1·5 × 15 × 15 | 194 gms. 1·5 × 15 × 15 | 88 gms. 1·5 × 15 × 15 | Square flake propellant. |

## ITALIAN 105 MM. 28 CAL. H.E. SHELL

(Fig. 32)

This shell is used in the 105 mm. 28 calibre gun-howitzer.

### Shell

The shell body is of forged steel the cavity being of the parallel wall type. Both cavity and exterior are machined over all. The cavity is threaded internally at the top to receive the nose adapter. The latter is a machined forging internally threaded at the nose to receive the fuze hole adapter, and externally at the base for attachment to the shell body. The fuze hole adapter is machined from bar stock and is threaded internally to receive the fuze and gaine and externally for attachment to the nose adapter.

The driving band is pure copper and a soldered steel base plate is fitted.

The overall length of the fuzed and filled shell is 16 inches approximately and the weight 35 lb. The weight of the empty shell is 28·4 lb.

### Bursting Charge

The bursting charge consists of 3 lb. $9\frac{1}{2}$ ozs. cast T.N.T. with central cast cavity for the exploder system. The surface of the T.N.T. is sealed with a plastic composition and the latter is covered with a felt washer.

### Exploder

The 43 gram ballistite exploder, screwed to the 5 gram ballistite gaine is similar to that described for the 149 mm./13 cal. high capacity H.E. shell in Pamphlet No. 9, page 31.

### Fuze

The Inneschi graze fuze used in this shell has been fully described in Pamphlet No. 7.

## ITALIAN 105/28 H.E. SHELL

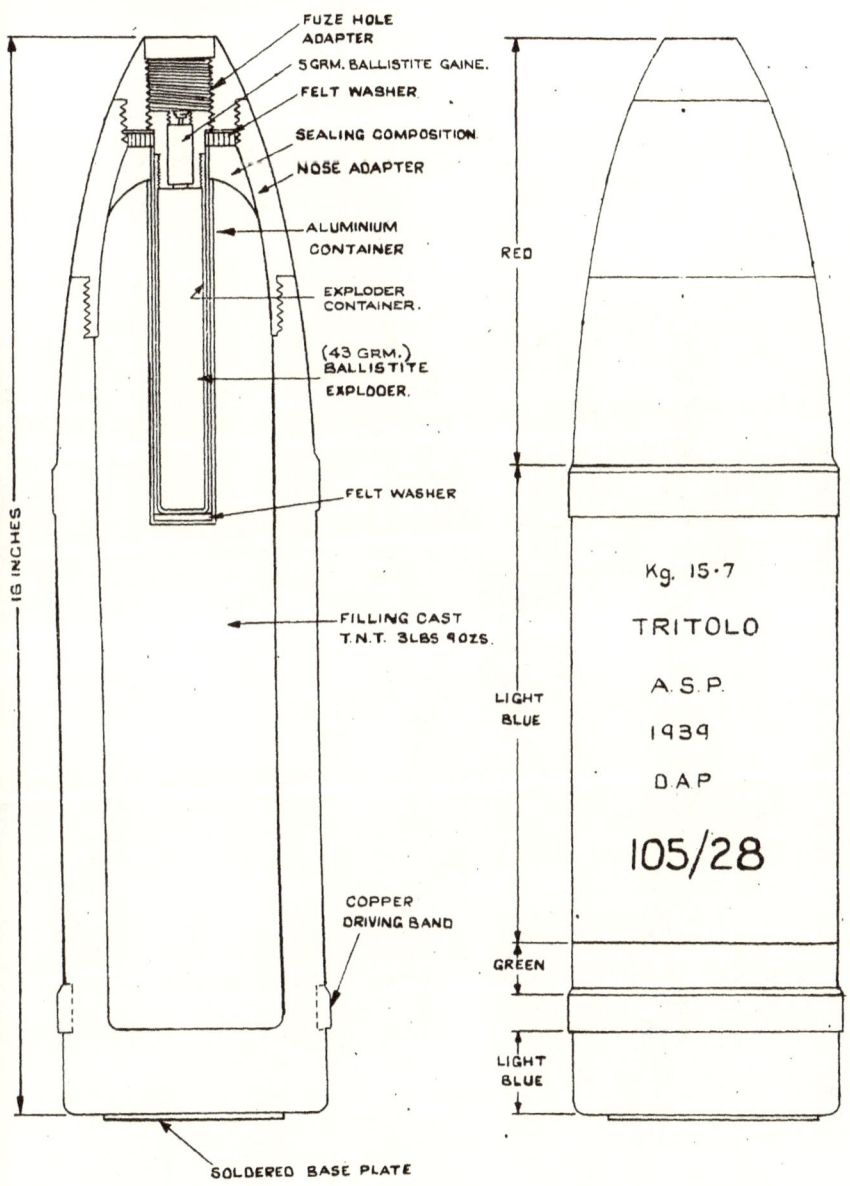

Fig. 32

## ITALIAN 75 MM. AMMUNITION TABULATION

This Table summarizes 75 mm. shell encountered for the equipments shown in the first column.

| Nature of Gun | Type of Round | Colour Markings of Shell | Stencil Markings of Shell | Method of Manufacture of Shell | Nature of Filling | Exploder System | Fuze and Gaine | Remarks |
|---|---|---|---|---|---|---|---|---|
| (A) 27 Cal. Gun, How. | Q.F. Separate H.E. | Head—Red Body—Light Blue Band above dvg. band—Green | TRITOLO 75/27 | Forged steel body, screwed head. Pressed in base plate. Streamline Copper dvg. band | Cast T.N.T. | Pressed T.N.T. Pellets in exploder container | Inneschi graze fuze, cyclonite gaine | Fully described in Pamphlet No. 7. |
| (B) 27 Cal. Gun, How. 13 Cal. How. | Q.F. Separate H.E. | Head—Red Body—Light Blue Band above dvg. band—Green | TRITOLO 75/13/27 | Machined from bar stock, screwed fuze hole adapter, soldered base plate. Copper dvg. band | Cast T.N.T. | Ballistite flake in metal tube | Inneschi graze fuze. Ballistite gaine | Fully described in Pamphlet No. 7. Weight filled and fuzed, 13 lb. 11 oz. 8 drs. |
| (C) 27 Cal. Gun How. | Q.F. Separate Hollow Charge | Cap—Red Body—Light Blue Band above dvg. band—White | 75/27 TRITOLITE | Forged Steel. Pressed in base plate. Aluminium alloy cap. Copper dvg. band. | Cast T.N.T./ Wax Hollow Charge | Cyclonite/wax in Aluminium sheath | Base fuze and gaine | Described in Pamphlet 8, details of fuze not yet available. |
| (D) 27 Cal. Gun, How. 13 Cal. How. | Q.F. Separate H.E. | Nose—Orange Body—Light Blue "Sheardized" metal coating. Band above dvg. band—Green over Yellow | Kg.6.2 TRITOLO A.P., F.M. D.A.Vr.XII./ 933 75/13/27 3 | Machined from bar stock. Screwed nose adapter. Soldered base plate. Copper dvg. band | Cast T.N.T. | Ballistite flake in metal capped cardboard tube, in Aluminium sheath | Inneschi graze fuze. Ballistite gaine | Identical in contour to (B) above. Weight filled and fuzed, 13 lb. 13 oz. 4 drs. |
| (E) 27 Cal. Gun How. | Q.F. Separate H.E. | Nose—Orange Body—Light Blue "Sheardized" metal coating. Band above dvg. band—Green. Rear of dvg. band—White. | Kg.6.100 TRITOLO D.A.Vr.935 75/27 | Machined as for (B) and (D) above | Detail not available | Detail not available | Detail not available | Identical in contour to (B) above. |

| | | | | | | | |
|---|---|---|---|---|---|---|---|
| (F) 27 Cal. A.A. | Q.F. Fixed Fragment- ation H.E. | Body—Blue | Kg.6.18 TRITOLO | Forged steel bodies. Machined fuze hole adapter of grey cast iron. Fragment- ation cylinder of machined grey cast iron. Copper dvg. band | T.N.T. cast in card- board cylinder | Ballistite flake in brass container | Time fuze of combustion type | One model fully des- cribed in Pamphlet No. 9. |
| | | Body—Blue Band—Dark Blue | Kg.6.33 TRITOLO A.P. | | | | | |
| | | Body—Blue Band—Dark Blue | Kg.8.1 or 8.2. TRITOLO A.P. | | | | | |
| | | Body—Light Blue Band—Dark Blue All types— Head—Orange | Kg.8.2. or 6.2 or 6.3 or A.P. 1928 TRITOLO A.P. | | | | | |
| (G) 27 Cal. Gun How. | Q.F. Separate Shrapnel | Nose—Orange Body—Red | Kg.6.6 75/27 | Forged steel body. Machined adapter and dia- phragm. Brass flash tube | Shrapnel similar in all respects to 77/28 Shell described in this pamphlet | As for 77/28 Shrapnel | Time fuze graduated to 59 (Pamphlet No. 9) | Comparable in all respects with 77/28 Shrapnel described in this pamphlet. |
| (H) 46 Cal. A.A. | Q.F. Fixed H.E. | Head—Red Body—Light Blue "Sherardized" metal finish. Bands—Blue Green above dvg. band | Kg.6.550 TRITOLO COMPRESSO 75/46 | Forged steel body. Pressed - in base plate. Machined nose fuze adapter. Double copper dvg. band | Not avail- able | Exploder container. Other details not yet available | No details available | Further details awaited |

NOTE.—Where the colour of the head of the shell is described in the third column as orange, this is due to a variation in the normal shade of the red used. The variation has no significance.

## JAPANESE 25 MM. CARTRIDGE Q.F. N.E/T.
### (Fig. 33)

This fixed Q.F. ammunition is fired by the Japanese 25 mm. A.A./A. Tk. gun. The round corresponds roughly to the British Hispano or Oerlikon types. The complete round weighs 8 oz. 10 drs. and comprises the following parts :—

    D.A. Fuze (Fig. 34).
    Shell and filling.
    Cartridge case and propellant.
    Primer (Fig. 35).

**Shell and Filling**

The streamlined shell is machined from steel bar and is coloured as indicated in Fig. 33. At the forward end it has a drilled cavity 1·89 inches in depth and 0·64 inch in diameter. This carries the bursting charge, and is threaded internally at the nose to receive the fuze. This cavity is coated with black enamel. At the base a tracer cavity 1·38 inches in depth and 0·47 inch in diameter is drilled. The cavity is closed by a rolled-in steel washer. Two copper driving bands are fitted. The rearward narrow band is 0·23 inch and the forward broad band is 0·69 inch wide. The leading element of the forward band is sloped gradually toward the nose of the shell.

The bursting charge, of total weight 5 drs., consists of 3 pellets of T.N.T./Aluminium composition (61/39 approximately). The upper pellet is annular and surrounds the magazine of the fuze. The whole is covered by a cardboard washer.

The tracing composition is of the following composition :—

| | |
|---|---|
| Magnesium | 16·5 per cent. |
| Wax | 3·2 ,, |
| Barium oxalate | 13·5 ,, |
| Barium chloride | 10·6 ,, |
| Sodium nitrate | 24·1 ,, |

Pressed over the top of the tracing composition is a priming layer of the following approximate formula :—

| | |
|---|---|
| Magnesium | 16·5 per cent. |
| Wax | 3·2 ,, |
| Barium oxide and peroxide | 55·8 ,, |
| Sodium nitrate | 18·3 ,, |

The total weight of tracing and priming is 5·25 drs. The surface of the priming composition has a deep conical indentation. The trace burns with an intense white light, the duration is, however, not known.

JAPANESE 25 mm., CARTRIDGE, Q.F., H.E./T.

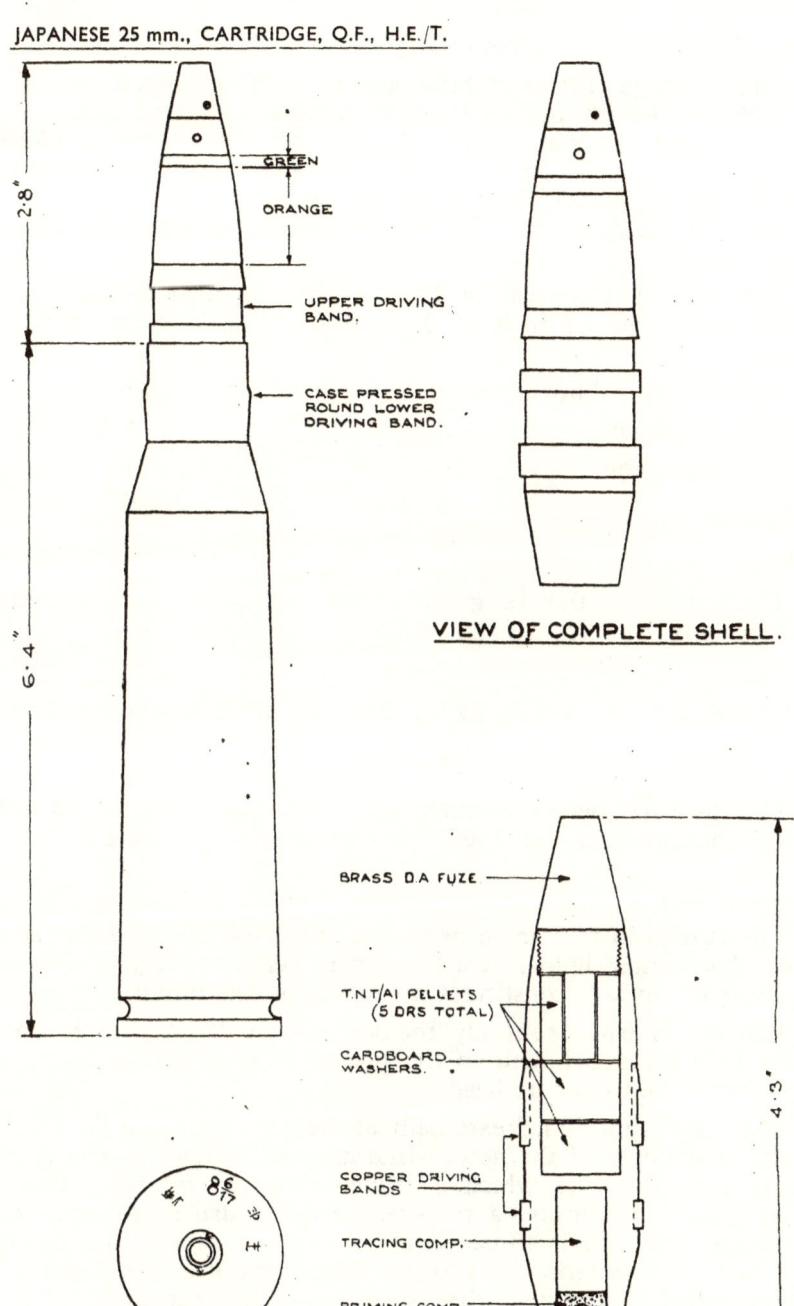

FIG. 33

## Cartridge Case and Propellant

The cartridge case is of brass and is of the rimless type. It is 6·4 inches in length and 1·67 inches in diameter at the base. It is heavily necked and is attached to the shell by rolling round the lower driving band with the lip of the case in contact with the rear of the forward driving band. The markings on the base of the case are indicated in the diagram. The marking 8·6/17 indicates the date as June, 1942.

The propellant consists of 3 oz. 10 drs. chopped tubular grains (approximate size 0·2 inch × 0·1 inch/0·02 inch) of the following composition :—

Nitrocellulose (Nitrogen content not determined)   92·9 per cent.
Dinitrotoluene   ...   ...   ...   ...   ...   5·7   ,,
Diphenylamine   ...   ...   ...   ...   ...   1·1   ,,
Graphite   ...   ...   ...   ...   ...   ...   0·3   ,,

## Primer

The primer is the 12 grain primer percussion Q.F. cartridge described on page 75.

## JAPANESE D.A. NOSE FUZE FOR 25 MM. AMMUNITION

(Fig. 34)

This fuze has been encountered in connection with the 25 mm. H.E./T. ammunition described elsewhere in this pamphlet.

## Construction

The fuze is in four main parts—the nose and body which are of brass, the shutter holder which is of nickel-plated brass, and the magazine which also constitutes a gaine is of duralumin.

The nose is drilled centrally, the bore being enlarged and threaded at the base for attachment to the body. It is closed by a copper diaphragm rolled into the head.

The body is machined externally at the top to provide the thread for the attachment of the nose. Internally an annular groove is cut leaving a central boss which is drilled through and constitutes a striker guide. Centrally, a transverse hole is drilled to carry the centrifugal bolt. It is threaded at one end to receive the closing plug, and has a centering boss for the bolt spring at is closed end. A hole is drilled centrally on the lower axis of this compartment and provides a channel for the striker needle. The plug is secured by stabbing and covered* with white paint. The centrifugal bolt is cylindrical, having a depression at one end to carry the retaining

## JAPANESE D.A. NOSE FUZE FOR 25 mm. SHELL

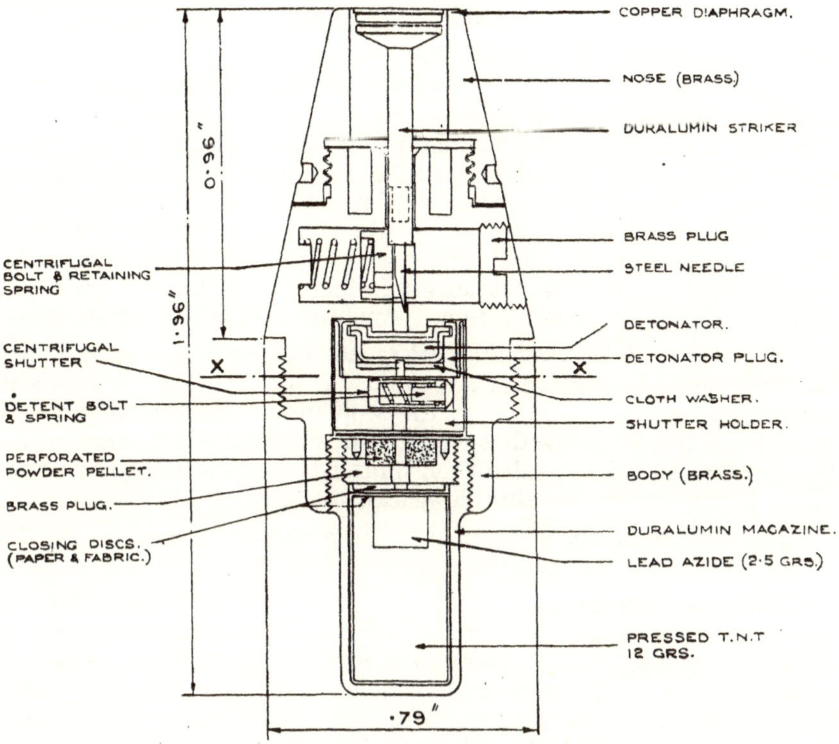

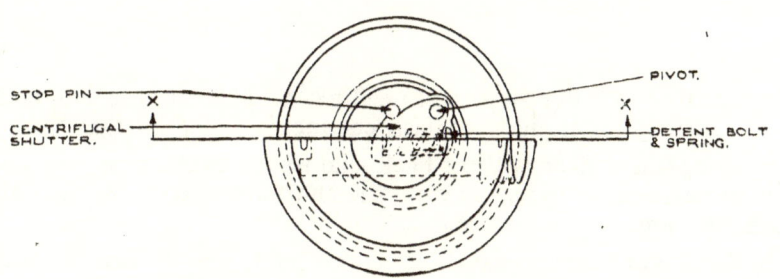

SECTION SHOWING SHUTTER & DETENT.

FIG. 34

spring and being stepped and slotted at the other to carry the end of the striker shaft and the steel needle. The striker is a headed duralumin shaft carrying a blued steel needle. This has the typical Japanese wedge-shaped joint. No support spring is fitted to the striker. At the base, the body is bored to receive the shutter holder, and tapped internally to receive the magazine.

The shutter holder is in the form of a cup fitting in the upper portion of the base cavity of the fuze. It has a widened bore at the top and carries therein the detonator plug. At the bottom it is drilled centrally to provide a flash channel, and carries the shutter pivot and stop pin. The detonator plug is of nickel-plated brass and retains the detonator by rolling the rim of the former on to the rim of the latter. The detonator itself comprises a copper shell and closing disc containing 0·93 grs. of a Mercury fulminate, Potassium chlorate, and Antimony sulphide, composition. (Analysis figures are not available).

The shutter is an oval brass plate notched on one side to engage the stop pin, and pivoted at one end. A blind hole is drilled transversely in the side opposite to the stop pin. The detent spring is inserted in this hole together with the detent bolt. The head of this bolt presses against the side of the shutter holder and the compression of the spring retains the shutter against the stop.

The shutter holder is retained in the body of the fuze by insertion of the magazine and suitable washers.

The magazine-gaine is of duralumin and is threaded externally for attachment to the base of the fuze and internally to receive the brass closing plug. The latter carries at the top an 0·93 gr. perforated powder pellet. The magazine contains a 12 gr. pellet of pressed T.N.T. with a central pressing of 2·5 grs. lead azide. This charge is covered by a series of washers as follows : one white paper, one open weave fabric, one perforated cardboard, and at the top white paper.

### Action

In the safe condition the striker needle is held clear of the detonator by the centrifugal bolt, and the flash channel is closed by the shutter.

On firing the striker will set back against the centrifugal bolt and lock it in position, at the same time the shutter sets back on the base of its holder and remains closed.

In flight the centrifugal bolt opens and the striker remains " floating " over the detonator, at the same time the shutter opens against the spring-loaded detent.

On impact the striker is driven on to the detonator, flash is boosted through the perforated powder pellet and actuates the magazine filling and so that of the shell.

# JAPANESE 12 Gr. PRIMER PERCUSSION Q.F. CARTRIDGE

(Fig. 35)

This primer is of the push-in type and fitted in the 25 mm. H.E./T. cartridge described elsewhere in this pamphlet.

## Construction

The primer body is of brass the base being 0·63 inch in diameter, and the length 0·67 inch. The body is slightly coned and is 0·49 inch in diameter at the top. It is unthreaded and is pressed into the primer boss. At the base it is drilled to provide a cap chamber, which is connected by a flash hole to a drilled magazine.

The copper cap is secured in the cap chamber by ring punching and contains 1·5 grs. detonating composition containing Mercury

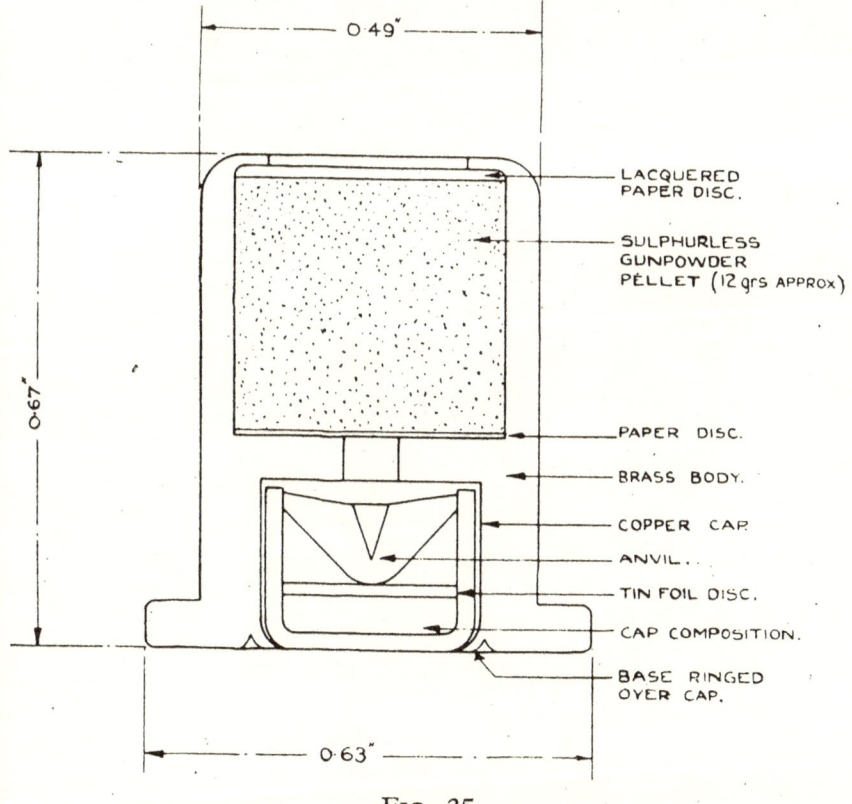

JAPANESE 12 GRAIN, PRIMER, PERCUSSION Q.F. CARTRIDGE

FIG. 35

fulminate, potassium chlorate and antimony sulphide. This is lacquered and covered with a tinfoil disc. The anvil is pressed into the top of this cap and is in contact with the detonating composition.

The magazine contains a pressed pellet of sulphurless gunpowder (potassium nitrate 85 per cent., charcoal 15 per cent.), and is covered by a lacquered paper disc. The top of the flash hole is covered by a thin paper disc.

**Action**

The cap is struck centrally crushing the detonating composition. Flash passing through the flash hole fires the gunpowder pellet and so the propellant charge.

Gas sealing is effected only by the expansion of the primer into the boss and the expansion of the cap in the cap chamber.

www.ingramcontent.com/pod-product-compliance
Lightning Source LLC
Chambersburg PA
CBHW032011080426
42735CB00007B/567